The Evol
of Electrical E

Books of Related Interest from IEEE Press

SPARKS OF GENIUS: Portraits of Electrical Engineering Excellence
By Frederik Nebeker, *IEEE Center for the History of Electrical Engineering*

This biography chronicles the ground-breaking technological advancements made by eight of this century's most brilliant engineers.

1994 Hardcover 280 pp IEEE Order Number PC0382-2 ISBN 0-7803-1033-0

TECHNOLOGICAL COMPETITIVENESS: Contemporary and Historical Perspectives on the Electrical, Electronics, and Computer Industries
Edited by William Aspray, *IEEE Center for the History of Electrical Engineering*

Written by world-renowned scholars, this is the first book to examine issues of technological competitiveness from a historical perspective.

1993 Hardcover 384 pp IEEE Order Number PC0324-4 ISBN 0-7803-0427-6

CLAUDE ELWOOD SHANNON: Collected Papers
Edited by N. J. A. Sloane and A. D. Wyner, *AT&T Bell Labs*
Sponsored by the IEEE Information Theory Society

The first published collection of papers by Claude Elwood Shannon, this book contains his seminal article "The Mathematical Theory of Communication."

1993 Hardcover 968 pp IEEE Order Number PC0331-9 ISBN 0-7803-0434-9

The Evolution
of Electrical Engineering

A Personal Perspective

Ernst Weber

President Emeritus
Polytechnic University

with
Frederik Nebeker

IEEE Center for the History of Electrical Engineering

IEEE
PRESS

The Institute of Electrical and Electronics Engineers, Inc., New York

© 1994 by the Institute of Electrical and Electronics Engineers, Inc.
345 East 47th Street, New York, NY 10017-2394

Printed in the United States of America

10   9   8   7   6   5   4   3   2   1

ISBN 0-7803-1066-7
IEEE Order Number: PP0420-0

Library of Congress Cataloging-in-Publication Data

Weber, Ernst (date)
    The evolution of electrical engineering : a personal perspective /
Ernst Weber, with Frederik Nebeker.
    p.   cm.
    ISBN 0-7803-1066-7
    1. Electric engineering–History. 2. Electric engineers–
Biography. 3. Weber, Ernst (date). I. Nebeker, Frederik.
II. Title.
TK15.W42   1994
621.3'09–dc20
                                    94-3165
                                    CIP

Contents

Preface

When I reflect on the history of electrical engineering, two factors stand out: electrical engineering evolved out of the science of physics and has always retained a close relationship with physics, and the theoretical formulation of the problems arising in electrical engineering has required higher mathematics and, in fact, has often stimulated new mathematical methodology.

Our fascination with electricity had already begun in ancient Greece, when people produced electric charges by rubbing amber and marveled at the attractive power of lodestone. But electromagnetic phenomena were hardly more than curiosities until the nineteenth century. In 1800 Alessandro Volta showed how to produce an electric current; in 1820 Hans Christian Oersted discovered that currents produced magnetic force; and in 1831 Michael Faraday found how to achieve the converse (inducing electric currents by means of magnetic forces).

There followed a period of empirical inventions and successful practical applications of electricity, such as the telegraph, electroplating, and arc lighting. Improvement in the means to generate current and in the design of an incandescent lamp led to Thomas Edison's central power station on Pearl Street in New York City in 1882. Edison used low voltage direct current. With the development of the means, especially efficient transformers, to produce and distribute alternating current, the electrical age burst into bloom.

The successful exploitation of alternating current owed much to Oliver Heaviside's mathematical formulation of the alternating-current theory, which was based on James Clerk Maxwell's epochal theory of electromagnetism. Maxwell's theory predicted the existence of electromagnetic waves, which would travel at the speed of light. In the late 1880s Heinrich Hertz demonstrated the generation, detection, and measurement of these waves, and within two decades Guglielmo Marconi and others had shown that these waves could be used to transmit messages over long distances without wires. Other momentous events of the end of the century were the discovery of the electron by J. J. Thomson and the development of electron beam devices, including the x-ray machine and the oscilloscope.

To illustrate the close connection between electrical engineering and mathematical physics, let us briefly consider the following examples, each of which will be described more fully later.

1. Applications of alternating current became effective with the introduction of the complex number notation, promulgated in the United States by Arthur Edwin Kennelly and Charles Proteus Steinmetz. These two engineers also endorsed the AC concept of impedance, introduced by Oliver Heaviside in England in 1884.

2. Long-distance telephony was greatly advanced by the introduction of loading coils, and this innovation resulted from the theory-based work of Michael I. Pupin and George Campbell, both of whom were much indebted to Heaviside's pioneering work on the subject.

3. The feasibility of communication networks spanning the frequency range from audio to microwaves was greatly advanced by the theoretical demonstration of frequency selectivity by wave filters, work done in 1915 both by George Campbell in the United States and by Karl Willy Wagner in Germany. The introduction of carrier frequency transmission depended upon wave filters.

4. The new world of long-distance, instantaneous communication by electromagnetic waves resulted from the genius of James Clerk Maxwell, who formulated the mathematical theory of electromagnetism.

5. The theoretical analysis of the electron flow in vacuum tubes by Irving Langmuir in 1913 made possible much more effective tubes, which eventually resulted in the mass production of tubes and the proliferation of types of electron tubes.
6. The negative feedback principle, utilized by Harold Black in 1927 for the stabilization of amplifier circuits, has been identified as one of the principal control elements in biological systems and with the work of H. W. Bode became the sine qua non of electronic circuit design. Advances in electronic control systems in the first decades of the century led to Norbert Wiener's coining of the word "cybernetics" for the general study of control systems, both natural and human-made.
7. The entire field of linear electric-network problems was systematized through the use of complex function theory, as initiated by Heaviside. Later, other methods, notably the Fourier integral method and Laplace transform techniques, further enriched this field, which has, in fact, in many instances stimulated the development of mathematical methodology, which then found applications in other areas of physics and engineering.
8. The study of energy conservation in semiconductor phenomena contributed to many advances in the application of semiconductors in electronics.

Electrical engineering is a young field. Practical applications go back only about a century and a half. (Samuel Morse's Washington-to-Baltimore telegraph line was completed in 1844.) Professional electrical engineering education goes back just over 100 years. (The American Society for Engineering Education celebrated its centennial in 1993.) Yet electrical engineering matured remarkably rapidly, as a result of the prior development of mathematical techniques and some parts of the relevant physics. In the chapters that follow I first describe how I came to be an electrical engineer; then I sketch, in the bulk of the book, how electrical engineering had become the mature field I found it, concluding with an outline of my career as an engineer and educator.

I

Entering the Electrical Engineering Profession

I.1   Growing Up in Vienna

The Austro-Hungarian Empire was, at the turn of the century, a motley assembly of a dozen or so different ethnic domains with different languages and customs. At its head sat the Emperor Franz Joseph (see Figure I.1), who had come to the throne in 1848. A reserved, conscientious, and rigid man, Franz Joseph held to the view that monarchy required personal command of the army and the state bureaucracy. From early in his reign to the very day of his death, he spent most of each day reviewing official documents.

Franz Joseph also saw himself as patriarch to all his subjects. He frequently traveled, making appearances at provincial capitals and many other places throughout the empire. He talked with people of all social standings and held regular Thursday audiences, when anyone could address the emperor. My grandmother once thus gained an audience with the emperor. And he made a point of conversing with his subjects in their native languages, including German, Hungarian, Italian, Croatian, Polish, Czech, Slovak, and Romanian.

Franz Joseph seemed, however, permanently saddened by a series of personal tragedies: his brother Maximilian, after a brief period as ruler of Mexico, was executed by a firing squad; his only son, Crown Prince Rudolph, committed suicide; and his wife Elizabeth was assassinated by an anarchist. Throughout, he

Figure I.1 A depiction of the emperor of Austro-Hungary, Franz Joseph (1830–1916).

persisted in what he saw as his duty, the administration of his empire. His personal fortitude and devotion to duty had earned him the respect and even the affection of most of his subjects. As a youngster in Vienna I certainly shared these feelings. It was a thrill for me to wave to Franz Joseph in his carriage—I would run alongside as long as I could keep up—as he departed or approached Schönbrunn Palace, which was not far from where my family lived.

It was the general belief that Franz Joseph was the essential bond that held the empire together, and that as long as he reigned, Austria-Hungary would somehow continue to survive its many difficulties. In 1908 he celebrated the sixtieth anniversary of his reign with great festivities that attracted worldwide attention. Those festivities are among my earliest recollections.

Vienna, the capital of the empire, was then the sixth largest city in the world. It had attracted a galaxy of artists, writers, and scientists. In music there were Gustav Mahler, Arnold Schönberg, Anton von Webern, and Alban Berg. There were the painters Gustav Klimt, Egon Schiele, and Oskar Kokoschka. Vienna's literary figures included Hugo von Hoffmannsthal, Arthur Schnizler, and Stefan Zweig. Among its philosophers were Ludwig Wittgenstein and Moritz Schlick, and Sigmund Freud and Theodor Escherich were part of its medical community.

Europe had seen no major war since the Franco-Prussian War of 1870, when Bismarck consolidated the German Reich. There had gradually been a realignment of the major powers: Germany's victory over France and its rapid military and industrial growth prompted England to become friendlier with France and with Russia, and Austria—which had not cultivated military prowess except through very elegant uniforms—moved closer diplomatically to Germany.

The Balkan Peninsula, however, was rife with ethnic conflict barely held in check by the Ottoman Turks, who ruled most of the region. Besides four Slavic groups—Serbs, Croats, Slovenes, and Montenegrins—there were Albanians, Greeks, and Bulgarians. The Ottoman Empire, the Austro-Hungarian Empire, and the Russian Empire all felt that their interests were at stake in political changes in the Balkans.

In 1908 Austria annexed Bosnia, but Bosnia contained a substantial Serbian population, most of whom favored union with Serbia. On June 24, 1914 the heir to the Austrian crown, Archduke Franz Ferdinand, who was a nephew of Franz Joseph, visited Sarajevo, capital of Bosnia. At about 10:30 in the morning the archduke and his wife were both assassinated by a member of "The Black Hand," a Serbian nationalist movement. Austrian officials claimed that the Serbian government knew of the planned assassination and failed to take any action to stop it or to inform the Austrians. On July 23 the Austrian foreign minister delivered an ultimatum to Serbia, demanding suppression of anti-Austrian elements. Germany backed Austria, and Russia backed Serbia, and when Serbia's reply did not satisfy Austria, the latter declared war on July 28.

As I was then almost 13 years old—I was born on September 6, 1901—and living in Vienna, I recall the excitement of that time quite well. I remember that by 1912 or so, there was a general feeling that war between England and Germany was inevitable. Germany was growing too fast for British interests, and the British insisted that German imports be marked "Made in Germany," because "Made in England" was then the mark of quality. Germany resented this and envied the English colonial empire. Shortly after war broke out, I recall seeing my uncle, who was a colonel in the infantry, ride out of Vienna at the head of a regiment, aching to fight. He was killed. I remember, too, the long lists of dead and wounded that appeared in Austrian newspapers.

Austria was not prepared for the kind of war that it found itself involved in. Fairly soon, the school I attended was converted to a military hospital to care for the many sick and wounded soldiers. I then had to walk farther to a make-shift school building, staffed by people deemed unfit for military service. This was my first year of secondary school. I had chosen not the eight-year course of humanistic studies, which required Latin and Greek, but the seven-year course preparatory for technical study. There the emphasis was on mathematics and physics, but we also received six years of instruction in French and three years in English.

My father was an employee of the Austrian railway system, and one of his jobs was keeping track of the use of railway cars and calculating what other railways owed the Austrian railway when its cars were used by them. I remember helping him with that bookkeeping task in the first years of the war.

Until I was 9 years old we lived in an old, large apartment building south of the city center. The building had massive walls and huge double doors leading to a large courtyard, which was dominated by a linden tree encircled by a wooden bench. The apartments, however, had no plumbing, though we had shared access to tap water and toilet facilities. We then moved to a modern apartment building with running water, our own toilet, and—what seemed little short of miraculous—electric lighting. My parents had four other children, all daughters. (See Figure I.2.) The last was born in 1915, and caring for a baby in wartime proved to be difficult.

On November 21, 1916 the Emperor Franz Joseph died at the age of 86, and his successor, Emperor Charles, was installed. Living conditions were becoming more difficult, and in 1917 food rationing began. It was fortunate that my facility in learning physics and French gained me a job assisting a classmate in those subjects; his father owned and operated a bakery, and I took payment in bread.

Even before the war ended, the Austro-Hungarian Empire began to dissolve. An independent Czechoslovakia was estab-

Figure I.2 A photograph of two of my sisters and myself.

lished in August 1918. On November 12, 1918, the day after the armistice and the abdication of Emperor Charles, the German section of Austria was proclaimed a republic and at once began to defend its borders as the territorial claims of the successor states frequently clashed. The political situation in Austria vacillated between socialist and conservative parties, with communist takeover—as actually happened in Hungary—threatened. I remember well the great unrest; one frequently saw trucks carrying armed soldiers rushing through the streets and heard rifles discharging.

The worst food shortages were in the last year of the war and in the months just after the armistice, as the allies did not lift the economic blockade of Austria until late March 1919. We went into the Vienna Woods to find anything green that could be used in a soup or salad. With the coming of food relief, principally by the United States and Great Britain, normal bread became available again. I remember that my youngest sister refused to eat it, as she had never known anything but black bread.

In February 1919 elections were held for a constituent assembly. This brought a semblance of order, and my school then returned to its original quarters. I attended my last year there and passed the final examination with honors on July 1, 1919, thus earning the *Maturitätszeugnis*, which entitled me to register for study at the technical university, then still designated *Technische Hochschule*. (See Figure I.3.) When I went for registration in the fall of 1919, I experienced a pleasant surprise: at the *Technische Hochschule* were American field kitchens that served us students hot cocoa and buns, which was for me at the time a wonderful delicacy.

Though my family suffered many hardships in these years, I was privileged in many ways. I had access to public libraries, and I read voraciously. Philosophy—especially the writings of Schopenhauer and Nietzsche—was a particular interest of mine. I experimented with electrical components and circuits, and I built a crystal radio. (I remember well instructing my parents in how to use the headphones; we listened to news and weather reports.) The interest in philosophy and the hobby of electrical experimentation do much to explain my decision to pursue both a humanistic and a technical education.

Figure I.3 A photograph of myself at about the time war broke out.

I.2 PURSUING A DUAL EDUCATION

Having completed *Gymnasium*, I decided to enroll in both of Vienna's universities. At the technical university (*Technische Hochschule Wien*, now *Technische Universität Wien*), I began the quite new program in electrical engineering. Actually, in those days the first two years were the same as the first two years of the mechanical engineering program. At the humanistic university (*Universität Wien*) I joined the division of philosophy, which included the sciences.

Three circumstances made it possible for me to pursue a dual education. First, the high school degree, the *Matura*, was recognized as qualification for enrollment at any university. Second, in Austria, as in most of Europe, education, including university education, was considered the government's obligation; professors were government employees, paid for by the state, and students paid only a small administrative fee. Third, classroom attendance was not required, and one earned a degree entirely by passing certain examinations.

These were exciting years for me. I had a great deal of freedom in choosing what lectures to attend, and I sought out the most stimulating professors. I had earlier taught myself shorthand—not an unusual skill for students to have in those days—and it helped in learning as much as possible from each lecture.

If I found unrewarding the lectures for a subject I would be examined in, I stopped attending and learned the subject on my own. The two universities were situated in different parts of the city, so I got plenty of exercise walking, and often running, from one to the other along the famous Ringstrasse.

These were exciting years in philosophy too. Indeed, one spoke of the "crisis in philosophy," which was in part a response to upheavals in physics. There was relativity theory—both Einstein's special theory dating from 1905 and his later general theory of relativity—and there was quantum theory. The latter originated in 1900 with Max Planck's study of light emission and developed through the efforts of Einstein, Niels Bohr, Erwin Schrödinger (in Vienna), and others. The new physics led to the questioning of fundamental concepts of ontology and epistemology.

After the first two years of study, I put my name down for the "First State Examination" on the topics I had studied at the technical university. Even though I had given more of my time to classes at the other university, I received on July 20, 1921, a certificate stating that I had passed with an overall rating of very good, with individual grades ranging from good to excellent.

Economic conditions were still very bad. Because the small postwar state of Austria was expected to pay the war debt of the entire Empire, national bankruptcy threatened. In late 1920 there commenced a severe inflation of the currency. I remember that in the summer of 1921, which I spent surveying in the high Alps east of Innsbruck for potential sites for hydroelectric power plants, we had to receive our earnings at least twice per week. As soon as we did, we ran to purchase food, clothing, or whatever we needed before the money lost much of its value. Early in the following year the value of one crown (*krone*) was down to about 1/25,000th of its earlier value, so you needed knapsacks to carry enough paper money for ordinary purchases. The inflation was brought under control after the conservative Dr. Ignaz Seipel was elected chancellor on May 31, 1922; he succeeded in arranging a large loan from the League of Nations that permitted the establishment of a new currency.

In early 1924 I felt ready to take the "Second State Examination" for a degree in electrical engineering. This consisted of a practical part and a theoretical part. For the former I was asked to design an electric power station with given specifications; this I did over a six-day period in late March. The latter was an oral examination, which I took on April 15. Again I passed with very good grades (see Figure I.4) and obtained the right to use the title *Diplom Ingenieur*.

I was now anxious to get employment, but realized that jobs were very scarce. I went to the office of the professor who taught electrical machine design and requested his recommendation. I had attended only a few of his lectures, which I had found to be not very illuminating, but in the oral examination I had expanded on his questions, so I believed that he had a favorable impression of me. He telephoned his friend Dr. Kann, the head of the personnel division at the Austrian Siemens-Schuckert Company. Emphasizing that Siemens was not at all looking for staff, he indicated I should visit Dr. Kann anyway and gave me directions to get to the Siemens plant in the outskirts of Vienna.

In the interview with Dr. Kann, I stressed that in addition to the courses at the technical university I had taken courses in mathematics and physics at the humanistic university. I mentioned too that I had already submitted two papers for publication. Kann spoke on the telephone with the chief of the design group and then sent me upstairs to visit him. After we had talked for a few minutes, he posed a problem, indicating that if I could solve it, I would be hired. His design group had not been able to explain why tests on a particular generator showed lower efficiency than they had expected. I examined the structure of the generator in detail and found that the pole shoes had a solid rather than a laminated construction, which, I knew, must cause appreciable eddy-current losses. Thus I earned employment at the Siemens-Schuckert Company, and I was thereafter frequently assigned special problems.

With a definite income assured, on January 18, 1925, I married Irma Lintner, who came from the same neighborhood as I did. We moved into a small apartment not far from either of our families. We had both become members of the Helvetic Confession, an undogmatic church that traces its roots to the sixteenth-century reformer Ulrich Zwingli. Austria under Emperor Franz Joseph was a strongly Roman Catholic country, and as a youth I was quite committed to Catholicism. Students were required to take religious instruction, and during the first two or three years

Akt.-Nr. 6069

TECHNISCHE HOCHSCHULE IN WIEN.

Prüfungsprotokoll-Nummer 689 v. J. 1924

Maschinenbauschule.
Unterabteilung für Elektrotechnik.

STAATSPRÜFUNGSZEUGNIS

(Zweite Staatsprüfung.)

Herr *Weber Ernst*

geboren im Jahre 1901 zu *Wien* in *Nieder-Österreich*

hat in den Studienjahren *1919/20 bis 1922/23*

die Technische Hochschule in Wien als ordentlicher Hörer besucht und in der Zeit

vom 24. bis 29. März 1924

die praktische Prüfung und am heutigen Tage vor der genannten Prüfungskommission die mündliche Prüfung abgelegt.

Auf Grund der umstehend im Auszuge aus dem Prüfungsprotokolle zusammengestellten Erfolge hat der Kandidat gemäß der Verordnung des bestandenen k. k. Ministeriums für Kultus und Unterricht vom 24. März 1912, RGBl. Nr. 59,

die zweite Staatsprüfung aus **Elektrotechnik**

sehr gut bestanden

und ist sonach auf Grund des § 10 der Kaiserlichen Verordnung vom 14. März 1917, RGBl. Nr. 130, zur Führung der Standesbezeichnung „Ingenieur" (Ing.) berechtigt.

Wien, am 15. April 1924.

Die Staatsprüfungskommission.

Der Vorsitzende:

Die Prüfungskommissäre:

473/1919

Figure I.4 The diploma I received in 1924 from the technical university.

of secondary school I was even serving in the Church. But my own interest waned, and after Franz Joseph's death in 1916 the atmosphere changed entirely. With a letter from my father, I was allowed to be excused from religious instruction. When I gave that letter to the priest, who was the instructor, he turned white with surprise and dismay, partly because I was the best student in the class.

Despite working full time at Siemens, I continued my studies at the university. At the humanistic university I worked on a dissertation with the well-known, and quite authoritarian, physicist Felix Ehrenhaft. My task was to carry out a theoretical analysis, using Maxwell's theory of electromagnetism, of the color produced when light is refracted by submicroscopic particles. The purpose was to understand better some experimental results by Ehrenhaft that seemed inconsistent with Millikan's famous experiments on the electrical charge of the electron. Ehrenhaft, who had used the apparent color of the droplets as a measure of size, obtained on several occasions values of charge less than what Millikan had hypothesized was the smallest unit of charge, that of the electron. My study showed that two quite different droplet sizes could produce the same color, and, thus, agreement with Millikan's result was reestablished. Ehrenhaft was somewhat disappointed, since he had challenged Millikan's results, but he accepted the scientific fact.

This work involved a great deal of calculation, for which I used a slide rule and—what was my lifesaver—the function tables of Jahnke-Emde. The dissertation was accepted, and I felt ready to take the final examinations for the Ph.D. degree. A problem was that the university rules required proficiency in Latin for that degree. Because I chose the technical course of studies in secondary school, I had taken no courses in Latin. Although I had learned some Latin on my own, I did not wish to be examined on that subject. Fortunately, I was able to get an exemption from that requirement because I was already a *Diplom Ingenieur*. I passed the two-hour oral rigorosum in physics and mathematics (with honors) on October 15, 1925, and the one-hour oral rigorosum in philosophy on November 9, 1925, and thus obtained the Ph.D.

In 1926 I continued my studies at the technical university (see Figure I.5), writing a dissertation, related to my work at

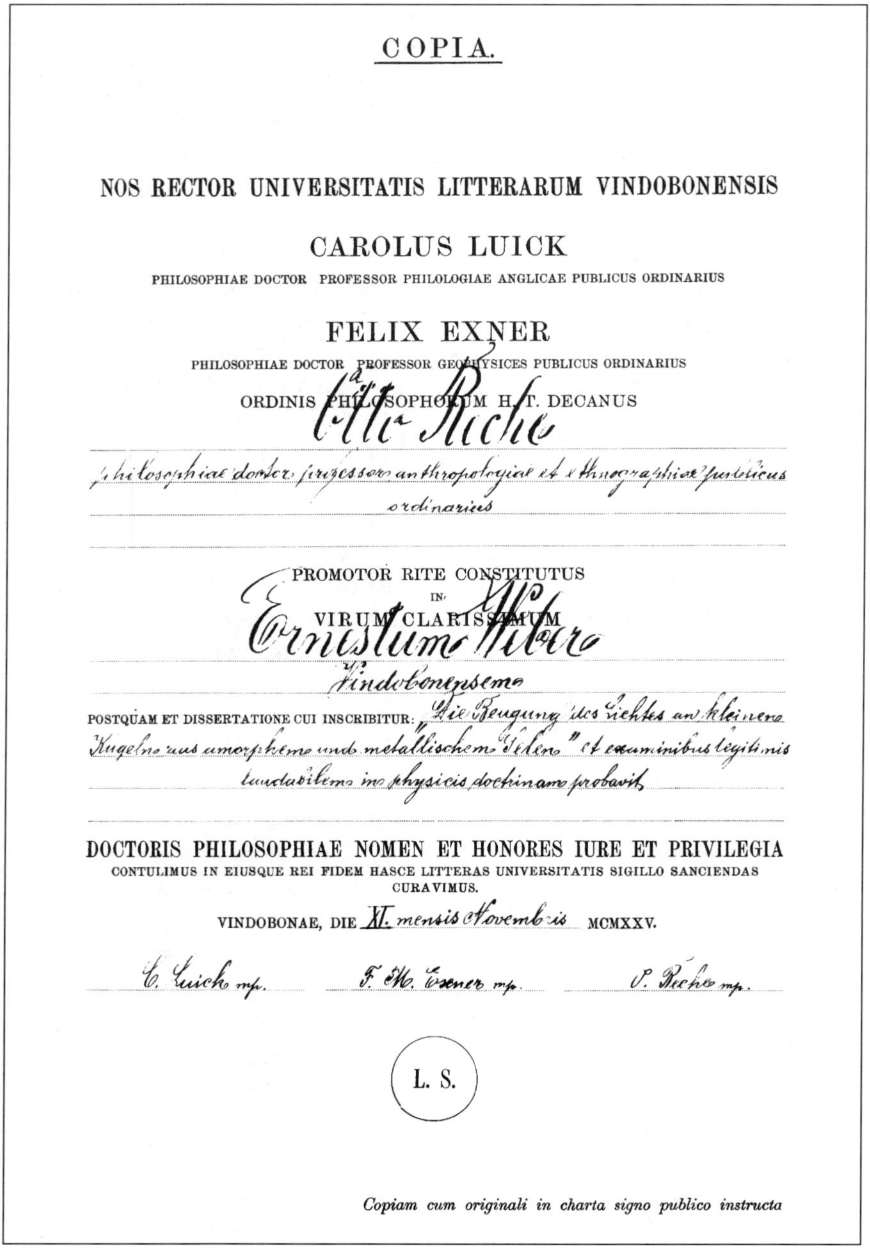

COPIA.

NOS RECTOR UNIVERSITATIS LITTERARUM VINDOBONENSIS

CAROLUS LUICK

PHILOSOPHIAE DOCTOR PROFESSOR PHILOLOGIAE ANGLICAE PUBLICUS ORDINARIUS

FELIX EXNER

PHILOSOPHIAE DOCTOR PROFESSOR GEOPHYSICES PUBLICUS ORDINARIUS

ORDINIS PHILOSOPHORUM H. T. DECANUS

Otto Reche

philosophiae doctor professor anthropologiae et ethnographiae publicus

ordinarius

PROMOTOR RITE CONSTITUTUS

IN

VIRUM CLARISSIMUM

Ernestum Weber

Vindobonensem

POSTQUAM ET DISSERTATIONE CUI INSCRIBITUR: *"Die Beugung des Lichtes an kleinen Kugeln aus amorphem und metallischem Selen"* *et examinibus legitimis laudabilem in physicis doctrinam probavit*

DOCTORIS PHILOSOPHIAE NOMEN ET HONORES IURE ET PRIVILEGIA

CONTULIMUS IN EIUSQUE REI FIDEM HASCE LITTERAS UNIVERSITATIS SIGILLO SANCIENDAS CURAVIMUS.

VINDOBONAE, DIE XI. *mensis Novembris* MCMXXV.

C. Luick mp.   *F. M. Exner* mp.   *O. Reche* mp.

(L. S.)

Copiam cum originali in charta signo publico instructa

Figure I.5 The doctoral diplomas I received from *Universität Wien* and *Technische Hochschule Wien*.

Abschrift.

Kraft des den Technischen Hochschulen erteilten Rechtes verleiht die

TECHNISCHE HOCHSCHULE zu WIEN

unter dem Rektorate des

v. ö. Professor, Hofrates Ing. Jngo Seidler

dem Herrn   *Ing. Dr. phil. Ernst Weber*

aus   *Wien*

den Titel und die Würde eines

DOKTORS DER TECHNISCHEN WISSENSCHAFTEN

samt allen damit verbundenen Rechten, nachdem derselbe im vorgeschriebenen Wege durch die von ihm vorgelegte Dissertation: "*Die magnetischen Felder und die Eisenverluste in leerlaufenden Synchrongeneratoren.*"

sowie durch die bestandene strenge Prüfung seine wissenschaftliche Befähigung erwiesen hat.

Gegeben zu Wien, am   *9. Juli 1927.*

v. ö. Prof. Dr. Ludwig Moser e. h.
Ordnungsmäßig bestellter Promotor.

L. S.

v. ö. Prof. Ing. Jngo Seidler e. h.
dzt. Rektor.

v. ö. Prof. Ing. Franz List e. h.
dzt. Dekan.
der *Maschinenbau* - Schule.

Die Übereinstimmung mit dem

Figure I.5 *(continued.)*

Siemens, on electric and magnetic field distributions; like my dissertation for the other university, this was theoretical rather than experimental work. In 1927 I received the Sc.D. from the technical university.

By the time my formal education ended in 1927, I was confirmed in my choice of electrical engineering as a career. My work at Siemens-Schuckert was stimulating, as I found that my training in mathematics and physics assisted me greatly in solving practical problems. I perceived that electrical engineering was becoming more and more an applied science, and I began to wonder about the historical development of the science of electrical engineering. As a youngster I had been fascinated by history—though ancient history was what received most emphasis in school—and the university courses I had taken in physics and engineering very often described the historical roots of current theories. I continued to take an interest in the history of electrical engineering, and what follows in the next three parts of this book is a sketch of my picture of that development up to the time I entered the profession.

II

The Dawn of the Electrical Age

II.1   EARLY INVESTIGATIONS OF ELECTRICITY AND MAGNETISM

Two thousand years ago the Greeks were aware of the action of amber on chaff and the action of loadstone on pieces of iron, but the first detailed report of the differences of these two effects was made by Gerolamo Cardano (1501–1576), professor of medicine at the University of Pavia. A wide-ranging scholar, Cardano is today remembered principally as a mathematician. He contributed to what was then a competitive enterprise of solving cubic equations, and he earned a place in the history of mathematics for countenancing mathematical expressions involving the square root of a negative number. In the history of electricity he is known principally as the author of the 1550 treatise "De subtilitate" ("On subtlety"), which summarized knowledge about amber and emphasized differences between the action of amber (which acted on light matter only) and that of loadstone.

The Englishman William Gilbert (1544–1603), also trained as a physician, read Cardano's treatise and began exploring materials that acted like amber. He constructed an instrument, the "versorium," that responded to many materials that had been rubbed. (We would call it an electroscope, since it revealed the presence of electric charge.) Gilbert described his investigations in one of the landmark books of Western science, *De magnete*, published in 1600. (See Figure II.1.)

Figure II.1 A woodcut from Gilbert's *De magnete* that shows a smith forming a magnet by beating a hot iron bar held in a north-south orientation.

Gilbert treated both electrostatics and magnetostatics, and he distinguished clearly between the two. Indeed, he coined the terms "Electricks" and "Magneticks" for substances manifesting a power of attraction. He described many experiments, including some with a spherical loadstone. He argued that earth acts as a giant magnet, and he investigated the geomagnetic phenomena of declination (deviation from true north) and inclination (deviation from the horizontal).

De magnete was extremely influential. Niccolo Cabeo (1586–1650), a graduate of the Jesuit College of Ferrara, carried Gilbert's experiments further, finding new "Electricks" and pointing out that repulsion occurs with "Electricks" just as it does with "Magneticks."

Also influenced by *De magnete* was Otto von Guericke (1602–1686), who is famous today for his construction of an air pump and the demonstration, using two teams of horses, of the difficulty of separating two hemispheres when the space they enclose has been evacuated. Gilbert's picture of the earth as a magnet seems to have inspired von Guericke in 1663 to make a sphere of sulfur and to rotate it with a stationary leather pad pressed against the sphere. This was the first electrostatic generator, and with it von Guericke achieved fairly high voltages (generating sparks up to an inch long). His experiments, which he described in a 1672 publication, demonstrated that an electrified body attracted to the sphere was subsequently repelled, but if afterward it was touched by a hand, it would again be attracted. This suggested two different actions of electrification.

England was the setting for several other advances in understanding electricity. Robert Boyle (1627–1691), one of the founders of the Royal Society, improved on von Guericke's air pumps and showed in 1675 that electric charges could be transferred in a vacuum (and thus answered the question of whether air was the carrier of the electric attraction). Francis Hauksbee (ca. 1666–1713) found that charge could be generated effectively by rubbing a long glass tube with paper. [Somewhat later, Charles François de Cisternay du Fay (1698–1739) introduced the distinction between vitreous electricity—produced by rubbing glass—and resinous electricity—produced by rubbing amber.] Hauksbee then constructed a machine on this principle, and it became the most popular electrostatic generator. Using this machine, Hauksbee demonstrated that electric charge could produce light within a glass tube.

Beginning in 1729 Stephen Gray (ca. 1666–1736) carried out a variety of experiments using a similar glass tube rubbed with paper. His most influential discovery was that the "electric vertue" could be conveyed considerable distances. Using packthread or twine, supported by silk thread, he transferred an electric charge as far as 270 meters. On the basis of capacity to transfer charge, Gray classified materials as conductors or insulators. He also demonstrated electrification by influence, bringing a charged rod close to, but not touching, an object, thus inducing electrification.

Another significant advance occurred in two places almost simultaneously. Ewald Georg von Kleist (1700–1748), living in what is today western Poland, placed a nail in the stopper of a water-filled vessel. He electrified the nail and received a much stronger shock than he expected when he later touched it. He found that the vessel held a large quantity of charge—he got almost equally strong shocks several times in succession—and held the charge for long periods of time. In the late fall of 1745 Kleist communicated his findings to friends. Peter von Musschenbroek (1692–1761) at the University of Leyden, apparently independently, discovered the same effect. He communicated his discovery to René Antoine Ferchault de Réaumur (1683–1757) early in 1746, and the device became known as the Leyden jar when Abbé Jean Antoine de Nollet (1700–1770) read a report concerning it to the French Academy of Sciences. Abbé Nollet went on to perform a dramatic demonstration before the king of France, having the shock from a Leyden jar transmitted through a long series of guardsmen. John Bevis (1683–1757) found that coating the glass jar with a thin layer of lead made the Leyden jar even more effective.

Word of these experiments reached even the English colonies of North America. In Philadelphia Benjamin Franklin (1706–1790), together with several other members of the Library Company (founded by Franklin in 1730), decided to acquire some simple equipment and to examine this new science of electricity. On March 28, 1747, Franklin wrote a letter—the first of a long series—to Peter Collinson, fellow of the Royal Society, in which he acknowledged the receipt of a Leyden jar: "Sir, your present of an electric tube, with direction for using it, has put several of us on making electrical experiments, in which we have observed some particular phenomena that we look upon to be new...." Franklin promised to communicate whatever discoveries he might make.

In a letter of July 11, 1747, Franklin (see Figures II.2 and II.3) reported one of his first discoveries to Collinson: pointed bodies are wonderfully effective in "drawing off" and "throwing off" the electric fire. Franklin advanced a single-fluid theory of electricity, the two types of charge being explained as an excess and deficit of a single electric fluid (and he introduced the designation "positive" and "negative" for the two types of charge). In a letter October 19, 1752, Franklin suggested that a kite experiment might confirm that lightning is an electrical phenomenon. Franklin carried out the experiment—by good fortune he was not injured—and then suggested the use of lightning rods to protect buildings. *Poor Richard's Almanac* for 1753 gave a description of how to avoid lightning damage.

Figure II.2   A portrait of Benjamin Franklin.

Franklin, who lived for quite a few years in London (much of the time as agent for the Pennsylvania Assembly), became acquainted with Joseph Priestley (1733–1804) in 1764. They became good friends and exchanged ideas about electricity. Priestley made the observation that the electrical force, like the gravitational force, must obey an inverse square law, since electric charges reside on the surface of a solid metal sphere. Franklin encouraged Priestley to publish a general review, which Priestley did in 1767 when *The History and Present State of Electricity, with Original Experiments* appeared. A few years later, Priestley made chemistry, especially chemistry of the atmosphere, his principal concern; he achieved lasting fame when in 1774 he isolated the gases now known as carbon dioxide and oxygen. Hostility to his liberal religious and political views caused

Figure II.3 A photograph of an electrostatic generator used by Benjamin Franklin.

Priestley's emigration from England in 1794. He and his family moved to Pennsylvania, where he continued his theological and scientific work.

In the eighteenth century there were several attempts to measure exactly the force between charged particles at various distances, and thus to determine experimentally and precisely the electrical force law. John Robinson (1739–1805) constructed an electrometer with which he made careful measurements. He arrived at the dependence of force on distance as inversely proportional to the distance to the 2.06 power, but the four-volume treatise reporting his experiments was not published until 1822, well after his death. In the early 1770s Henry Cavendish (1731–1810) made careful measurements and arrived at a force law where the exponent was 2.02, but here again the results were not published until much later.

The person famous for measurements of electrical force is Charles Augustin de Coulomb (1736–1806). In 1785, using a delicate torsion balance devised for this purpose, Coulomb showed that the force of attraction between charges is, indeed, inversely proportional to the square of their distance. He went on to formulate a force law, with the attraction directly proportional to the product of the charges. He reported these results in the *Mémoires de l'Academie Royale des Sciences*. Coulomb also applied the same experimental arrangement to determine the force between magnetic poles, making corrections for the effect of the earth's magnetic field.

Thus by the late eighteenth century, investigators had learned much about electrostatics and magnetostatics and had even discovered a few quantitative laws. Yet much of the realm of electromagnetic phenomena was entirely unknown. The discovery of a means to sustain a current of electricity—described in the next section—opened up much more of that realm.

II.2   ELECTRIC CURRENTS AND OHM'S LAW

Though science and politics are different realms of human activity, they often are interrelated. An example is provided by the French Revolution. King Louis XVI, experiencing great financial problems, convened a national assembly in June 1789. There followed an uprising, the declaration of France as a republic, and the execution of the king. At this time the reaction against all traditional ways was so strong that Charles Maurice de Talleyrand could ask the national assembly to initiate studies for a new system of weights and measures. A group of a dozen distinguished scientists devised a rational, coherent system based on a unit of length (the meter, 1/10,000,000th of the earth's meridional quadrant), a unit of mass (the kilogram, the mass of a cube of water 1/10th of a meter on a side), and a unit of time (the second, 1/36,400th of the mean solar day). The national assembly hoped that the system would be accepted by other countries and sought to interest the British Parliament in a cooperative effort. They were unsuccessful in the latter, but gradually a number of other European countries adopted this so-called metric system. There was something of the same rationalizing spirit in the formation of the United States at about the same time. There a decimal monetary system was adopted, but otherwise English measures were retained.

The American Revolution played a part in the emigration of an outstanding scientist, Benjamin Thompson (1753–1814), later named Count Rumford. After living in England for some years, he entered in 1785 into the service of the elector of Bavaria. There he served as minister of war and conducted experiments that called into doubt the prevailing theory of heat as a fluid. Having returned to England, Rumford joined in 1799 with Sir Joseph Banks, president of the Royal Society, to establish the Royal Institution "for diffusing the knowledge and facilitating the general introduction of useful mechanical inventions and improvements, and for teaching by courses of philosophical lectures and experiments the application of Science to the common purpose of life." Humphry Davy (1778–1829) was selected as scientific lecturer in 1801 and designated professor in 1802. Davy did outstanding research in chemistry, and his lectures became famous. In the history of electrical technology he is notable for the discovery of arc lighting and for hiring, in 1813, the young Michael Faraday as assistant.

Another professor of the Royal Institution was Thomas Young (1773–1829). Although Newton's corpuscular theory of light was most widely accepted, Young and others accepted Huygens's concept of light as a wave phenomenon. Young argued that light waves were oscillations transverse to the direction of propagation, and he demonstrated diffraction color bands on knife edges.

Stronger evidence for the wave nature of light came from several French investigators. Étienne-Louis Malus (1775–1812) provided support in his studies of polarization upon reflection of light from crystal surfaces. Stronger and eventually incontrovertible evidence came from Augustus Jean Fresnel (1788–1877). Fresnel, trained at the engineering school *École des Ponts et Chaussées* (School of Bridges and Roads), mastered the analytical methods of Jean Baptiste Joseph Fourier (1768–1830) (presented in the 1822 *Traité de la chaleur*) and demonstrated mathematically how the observed diffraction bands could be explained as interference of light waves. Working with Dominique François Arago (1786–1853), Fresnel provided further evidence for the wave nature of light in studies of the interference between two rays of polarized light. When it later became possible to measure accurately the speed of light, the wave theory prediction was confirmed that the refractive index of a medium is, in fact, the ratio of the speed of light in vacuum to the speed of light in the medium. As we will see shortly, it was through the work of James Clerk Maxwell that this line of research merged with the study of electrical and magnetic phenomena.

Another research tradition that rather unexpectedly became linked to electrical science was animal physiology. Albrecht von Haller (1708–1777), a Swiss anatomist, studied the interaction of muscles and nerves, making a distinction between sensibility of tissue (that is, the feeling of sensation) and irritability (that is, muscular contraction in response to a stimulus conveyed by the nerve). The capacity of an electric shock, as delivered by a Leyden jar, to stimulate muscular contraction turned interest toward the electric eel and the torpedo fish, whose electric organs were studied.

At the University of Bologna in 1780, Luigi Galvani (1736–1798) demonstrated for his students the nerve-muscle relationship of a frog's leg and discovered that contact with a scalpel caused a sudden contraction of the muscle. Galvani believed that the scalpel triggered a release of "animal electricity," which in turn made the leg jerk. In 1791 he published a memorandum and sent a copy also to his friend Alessandro Volta (1745–1827) at the University of Pavia. Volta (see Figure II.4) repeated the experiments in many variations, and in 1797 he came to the conclusion that what Galvani had interpreted as "animal electricity" was in fact "contact electricity" of two dissimilar materials. Volta discovered that a stack of alternating zinc and copper disks, if kept moist, generated a constant electric current.

This was the invention of the electric battery (called initially a "voltaic pile"; see Figure II.5) and the beginning of a new phase of electrical experimentation—for the first time, a constant current was available. Volta had earlier made two other inventions of value to the new science of electricity: an alternative to fric-

Figure II.4   A portrait of Alessandro Volta (courtesy of the
Burndy Library).

tion electrostatic machines, the "electrophorus," which used
electrostatic induction to build up charge, and a condensing electroscope, with which very small electric charges could be detect-
ed. These inventions had won him election to the Royal Society
in 1791 and that Society's Copley Medal in 1794. Shortly after
Volta's invention of the battery, he described the device in a letter to Joseph Banks, president of the Royal Society, and it was
not long before people on both sides of the Atlantic were making
voltaic piles. In England, Anthony Carlisle (1768–1840) and William Nicholson (1753–1815) constructed a pile using disks of silver and zinc. They noticed gas bubbles around a wire dipping

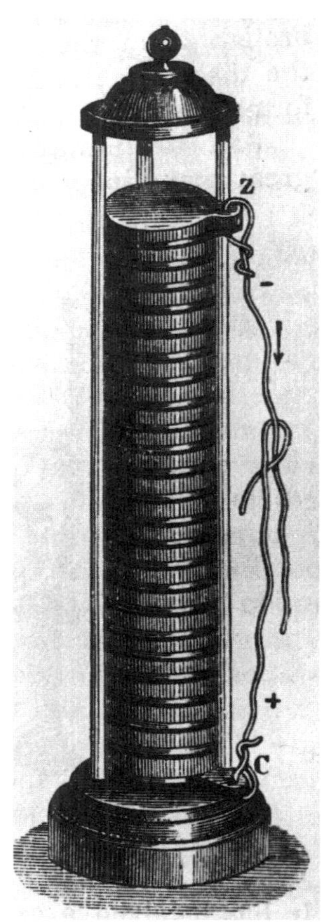

Figure II.5 An illustration of a voltaic pile.

into water and were able to identify the gases, oxygen and hydrogen, and thus witnessed the electrolysis of water.

Volta, in his letter to Joseph Banks, had mentioned the arrangement of individual cells or cups in series as a "crown of cups." At meetings in Paris in 1801, Volta was asked to demonstrate his pile before Napoleon, then first consul, and in recognition of his achievement he was awarded a gold medal. Later, as emperor, Napoleon presented to the *École Polytechnique* a giant battery of 600 pairs of "cups." In 1808 the Royal Institution in London acquired a giant battery of 2000 pairs, and it was with this battery that Humphry Davy in 1809 demonstrated a brilliant electric arc between carbon rods.

The new science of electricity was coming to be studied in universities. In 1779 Alessandro Volta, at the age of 34, was appointed to the newly created chair of physics at the University of Pavia. At the University of Copenhagen, Hans Christian Oersted (1777–1851), professor of physics, lectured regularly on both electricity and magnetism. Indeed, it was at a lecture demonstration that he first noticed that an electric current caused the deflection of a magnetic needle. He concluded that the current surrounded itself circularly with magnetic action and showed that a current loop produced magnetic action just as a magnetic pole did. Oersted provided, for the first time, clear evidence of the interrelationship of electricity and magnetism. And his report of the discovery, published in several languages including Latin (still a language of scholarship in northern Europe), stimulated experimentation in many countries. (See Figure II.6.)

When the news of Oersted's discovery reached Paris, Dominique François Jean Arago repeated the experiments, and on September 11, 1820, he demonstrated them before the French Academy of Sciences. Before the end of the following month, Jean Baptiste Biot (1774–1862) and Felix Savart (1791–1841), both professors of physics, were able to show that the force between a current-carrying wire and a magnet pole was inversely proportional to the distance between them. Arago's colleague, André Marie Ampère (1775–1836), professor of mathematics of the *École Polytechnique,* soon derived a mathematical theory of attraction and repulsion between two parallel wires.

Ampère's suggestion that this force could readily be used in an instrument to measure the electric current strength, by suspending a small magnet in a coil of wire, was taken up by Johan Saleno Christoph Schweigger (1779–1857), professor at the University of Halle, Germany. Ampère suggested "galvanometer" as the name for the instrument; because Schweigger used a coil instead of a single wire, he called it the "galvanic multiplier." Others soon constructed similar instruments, including Michael Faraday in London. In Berlin Hermann von Helmholtz developed the theory in order to permit calibration, though there was still lacking an agreed-upon standard of current strength.

Arago himself performed an experiment in 1824 that had several important consequences much later. He rotated a cop-

Figure II.6 An artist's depiction of Oersted's discovery that the current of a voltaic pile caused the deflection of a magnetic needle (courtesy of the Burndy Library).

per disk and observed motion of a magnet needle in proximity. When placed appropriately, the needle would begin rotating in the same direction as the disk. Though a mystery at the time, what Arago had observed was the induction of eddy currents in the disk by the magnetic field of the needle, followed by the action of the magnetic field of the currents on the magnet.

Arago also noticed that a piece of iron placed inside a conducting coil could be magnetized. This observation prompted Ampère to wind a wire into a helix to produce a strong magnet, which he called a "solenoid" (from the Greek word for "pipe shaped"). In 1825 William Sturgeon (1783–1850) in England and, at about the same time, Joseph Henry in the United States found that by winding wires around a soft iron core one could produce an electromagnet many times stronger than bar magnets.

However, the relation between the driving force of a voltaic pile (which soon came to be called "voltage") and the resulting current in a closed circuit was unknown; in fact, some doubted that they were related at all. In 1825, Georg Simon Ohm (1789–1854), professor of mathematics at the Jesuit College in Cologne, began to examine this relationship both theoretically and experimentally. In his theoretical investigations he was guided by the theory of heat flow published by Jean Baptiste Joseph Fourier in 1822. For his experimental investigations, he prepared his own metallic wire samples of different lengths, used boiling water and melting ice for constant temperature references, and assured control of current through a galvanometer, the only sensitive instrument available. Control of the electromotive force of the pile was crucial, and to achieve it, Ohm used the thermoelectric effect which had been discovered by Thomas Johan Seebeck (1770–1831) in 1822. Though the final publication of his exacting measurements in 1827 still elicited criticism, independent verification in following years led to full recognition of his fundamental contributions. In 1839 he was elected to membership in the Berlin Academy of Sciences, in 1841 he received the Copley Medal of the Royal Society of London, and the following year he was named a foreign member of the Royal Society. His most cherished recognition came in 1849 when he was appointed to the chair of physics and mathematics at the University of Munich. Today, of course, the linear relationship between current and voltage for metallic conduction is known as Ohm's law.

II.3 FARADAY'S EXPERIMENTS LINKING MAGNETISM AND ELECTRICITY

Michael Faraday (1791–1867) was the son of a blacksmith who struggled to make a living in the outskirts of London. His parents belonged to the Sandemanian Church, a sect somewhat similar to the Quakers. Faraday received little formal education and had to begin earning money at an early age. He was only 13 when he was apprenticed to a bookbinder, a French émigré who had fled the French Revolution and started a new life in England. Faraday not only bound books, but also read them voraciously. Isaac Watts's *Improvement of the Mind* made a deep impression on him.

In 1808, at age 17, Faraday started attending the City Philosophical Society meetings at which John Tatrun lectured on science topics. A few years later he attended a series of lectures given by the celebrated chemist Humphry Davy, professor at the Royal Institution. Faraday took careful notes, and later presented a bound copy of these notes to Davy. Davy, who had just injured his eyes during a chemical experiment, offered Faraday the position of laboratory assistant with full use of the equipment in two rooms on the top floor of the Institution. The Royal Institution was to remain Faraday's base of operations for his entire life.

In 1813 Davy arranged a tour of the Continent, through France, Italy, Switzerland, Germany, and Belgium. For Faraday, who accompanied Davy as secretary, this was a formative educational experience, and upon return to London in 1815, Davy made him superintendent of all apparatus in the chemical laboratory of the Royal Institution. Faraday became proficient at chemical experimentation; he made new compounds of chlorine and carbon and isolated benzene by fractional distillation of hydrocarbons.

When news came in 1820 of Oersted's observations of the magnetic action of an electric current, Davy wanted to see these experiments repeated. Oersted had, of course, used a voltaic pile to produce the electric current, and Davy already had much experience with that device. As mentioned earlier, in 1809 he used a giant voltaic pile to produce a brilliant electric arc between carbon rods.

Faraday was able to repeat Oersted's experiment without difficulty, and he at once reflected that if the magnetic needle is deflected by current, then by Newton's third law, a current should rotate around a magnet. By an ingenious arrangement, he was able to make a current-carrying wire rotate about a bar magnet. He was even able to see a delicately balanced current-carrying loop respond to the earth's magnetic field.

These demonstrations suggested that a practical electric motor might be constructed, and many inventors, such as Francis Watkins in London and Thomas Davenport in Vermont, made the attempt. In most instances, however, the contrivances were of the nature of toys. Joseph Henry (1797–1878) constructed an electric motor that imitated the reciprocal motion and the force action of the steam engine, but by means of magnets. Henry himself had no optimistic view of the future of such devices, principally because in order to drive electric motors one needed batteries—at that time there was no other power supply—which were costly, unreliable, and heavy. Enterprising craftsmen nevertheless carried on. Moritz Herman von Jacobi (1801–1875) of St. Petersburg, with a grant from the tzar, used batteries, an electric motor, and paddlewheels to move a boat on the Neva River in 1839, and in the following year Robert Davidson, helped by a grant from the Scottish Society of Arts, showed that a wheeled vehicle could be propelled by an electric motor.

Oersted's demonstration suggested to Faraday not only electric motors, but also electric generators. In an 1822 diary entry, Faraday reasoned that if an electric current can produce magnetic effects, then a magnet should produce electric effects, yet he was able to find no effect of a magnet on an electric circuit. Others, including Ampère, A. J. Fresnel, and Daniel Colladon, had the same idea and were likewise unable to demonstrate an effect. In retrospect one sees that these people were looking for an effect in a stationary system, therefore did not find one. Yet in arranging the experimental setup—moving either a coil or a magnet—they must have produced inductive effects. "To be prepared for the unexpected" is certainly an indispensable attitude for researchers.

Faraday's work had already made a deep impression upon the scientific community, so that he was made a member of the Royal Society in 1824, even against the wishes of his superior, Humphry Davy, who apparently feared becoming overshadowed by his assistant.

Faraday continued to seek a way to induce an electric current with magnets. In 1824 he repeated Arago's experiment revolving a copper disk between magnets, which seemed a step in the right direction, but he could not find the next step. In 1831 he tried a new experimental setup: a soft iron ring on which were wound two coils of copper wire, one connected through a switch to the battery and the other to a galvanometer (see Figure II.7). (His arrangement was, in fact, the prototype of a transformer, though he worked only with direct current.) When Faraday

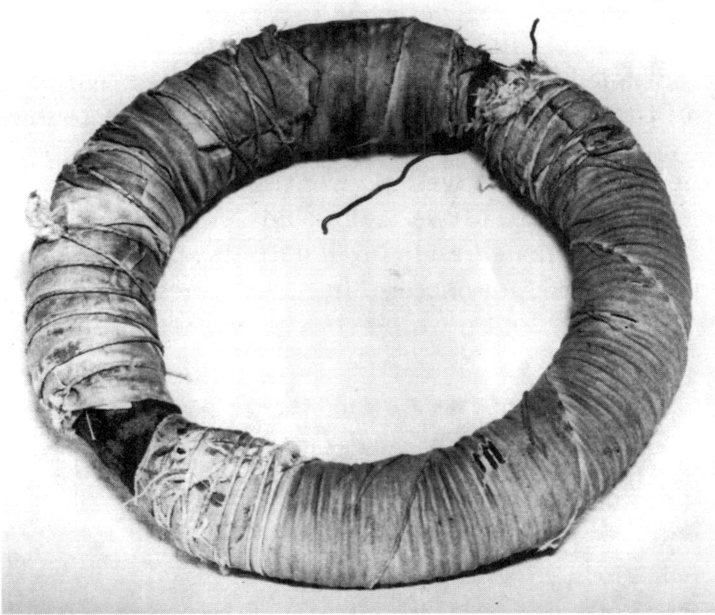

Figure II.7　A photograph of the induction ring used by Faraday (courtesy of the Burndy Library).

switched the battery on, he observed that the galvanometer jumped, indicating a momentary current in the second coil. He repeated this, finding that the galvanometer registered current only when the current was switched on or off. To be certain that the magnetic linkage was important, Faraday replaced the iron ring by a copper ring and found no noticeable effect upon the galvanometer.

Faraday then used a larger cylindrical solenoid, such as Ampère had constructed, connecting its wire ends to a galvanometer. The approach of a bar magnet caused a slight motion of the galvanometer needle. However, when he plunged the bar into the coil, the galvanometer made a very definite indication, as well as when he withdrew the bar magnet. Faraday thus discovered that relative motion, as well as a changing magnetic field, could generate an electric current.

With this effect as his cue, Faraday again performed Arago's experiment, this time using a copper disk about 12 inches in diameter and a very large magnet belonging to the Royal Society.

With the disk rotating between the magnetic poles, he used brushes at the axis and the rim of the disk to take off current. He found that a feeble direct current was generated (with this prototype of the homopolar generator). Faraday then replaced the disk by a simple wire "cutting the magnetic lines of force." The latter he visualized, as was his wont, by sprinkling iron filings on paper in the space between the poles.

With this confirmation in hand, Faraday (see Figure II.8), on November 24, 1831, announced his discoveries before the Royal Society, stressing that *a change in the magnetic field*, either by

Figure II.8   A portrait of Michael Faraday (courtesy of Argonne National Laboratory and the Niels Bohr Library).

motion or by variation of the field in time, would produce an electric current in a nearby wire loop. He called this effect "electromagnetic induction." His announcement caused a sensation, and at once many types of "electric generators" were produced in various countries. Their output, however, was so small that no significant practical application was found, and this remained the case for some two decades.

In the United States, Joseph Henry (1797–1878) had conducted similar experiments, motivated by similar reasoning, slightly before Faraday's work. However, with meager means and considerable teaching obligations, Henry moved rather slowly. Moreover, he wanted to be quite certain that his observations were repeatable before making any results public. He learned of Faraday's work when the *Philosophical Transactions* announcement reached him in May 1832. Though disappointed that he had not earlier published, Henry then submitted several papers to Yale College's Benjamin Silliman for publication in his journal, *The American Journal of Science*.

The first electromagnetic generator was constructed in 1832 by Hippolyte Pixii (1808–1835), a Parisian instrument maker. A permanent magnet in horseshoe form rotated about fixed coils and delivered current of alternating direction. Since the world was familiar only with direct current, Ampère invented a primitive commutator that produced undulating, but unidirectional current. A great improvement was provided by William Ritchie (1790–1837), who kept the large horseshoe magnet fixed and rotated the coils.

Investigators had long wondered about the nature of electricity and had questioned whether atmospheric electricity (as manifested in lightning), frictional electricity (often stored in Leyden jars), and chemical electricity (generated by voltaic piles) were the same. It was Faraday who, through a series of experiments, finally convinced most researchers that the various electricities were identical. The nature of electricity, though, even after the discovery of the electron and the success of quantum electrodynamics, remains enigmatic.

As assistant to Humphry Davy, Faraday became familiar with Davy's work in electrochemistry, notably the decomposition of the alkali salts potash and soda in 1807 and the isolation of the new elements that Davy called potassium and sodium. After

Davy's death in 1829 and Faraday's discovery of electromagnetic induction in 1831, Faraday turned to electrolysis. By 1833 he had established the basic quantitative relationship known today as the Faraday law of electrolysis: a specific amount of electricity liberates one gram-atom equivalent of substance from a solution capable of conducting electric current. Thus, the same amount of electricity will deposit 108 grams of silver or 23 grams of sodium, these being the relative gram-atom equivalents. Faraday, with the help of the classically trained scholar William Whewell, introduced terminology that has since become universal: electrode, electrolyte, anode, and cathode.

Faraday had observed that, with any circuit that included coils, making or breaking the circuit produced sparks. Finding that increasing the number of turns of the coil led to heavier sparks, Faraday concluded that the spark-producing current was a result of self-induction caused by the decay of the magnetic field. Independently, Joseph Henry arrived at the same result. Closely related is the 1834 conclusion by Heinrich Friedrich Emil Lenz (1804–1865) that every induced current delays the change of the magnetic field causing the induction, which is today known as Lenz's law.

The induction coil soon found use in the medical treatment known as "galvanism" or shock therapy. Charles Grafton Page (1812–1868) built a particularly effective induction coil, which he called the "dynamic multiplier": he placed a long coil on a soft iron core, fed a short section by battery, and provided several terminals along the remainder of the coil, thereby obtaining successively higher voltages and stronger shocks.

As research in electromagnetism progressed and as applications of electricity such as galvanism and telegraphy emerged, it became increasingly important to be able to specify strength of current. At first the only instruments available were galvanometers giving arbitrary measures. Early efforts to establish standard measures were related to the metric system introduced in France in 1795.

In the metric system, centimeter, gram, and second were the basic units (hence the system is generally referred to as the cgs system). Initially, of course, there was no provision for electrical measurements. It was Carl Friedrich Gauss (1777–1855) who first established a relationship. In 1833 he published the paper,

"On the intensity of the Earth's magnetic field expressed in absolute measure," measuring this field strength by means of a magnetometer in the cgs system of units.

Gauss was then in charge of the geomagnetic observatory in Göttingen. His emphasis upon absolute measurements of the new magnetic quantities in terms of the cgs system of units motivated Wilhelm Eduard Weber (1804–1891), who had joined him as co-director, to do the same for other electrical quantities. Help came in 1845 from Franz Ernst Neumann (1798–1895), who deduced the mathematical relation for the induced current in Faraday's experiment, as well as the general formula for mutual inductance between different circuits. Weber had already deduced the force acting between moving charges and applied this to evaluate the induced current in circuits. Making use of Neumann's work, he was able to determine the absolute units for current, electromotive force (voltage), and resistance, expressed in the fundamental cgs units of mass, length, and time. Weber also invented the electro-dynamometer, which measured the force between two current-carrying coils and which was more accurate than an instrument incorporating a permanent magnet.

One of the problems with absolute measurement that Weber and other physicists encountered was that two ways of measuring the charge on a capacitor—electrostatically (i.e., measuring the charge held) and electromagnetically (i.e., measuring the discharge of that capacitor)—gave quite different values. In 1856 Weber and Rudolf Kohlrausch (1809–1858) set out to measure this ratio, which had the physical dimension of velocity (when the two measurements were made in, respectively, cgs electrostatic units and cgs electromagnetic units). Weber and Kohlrausch found the value to be about 3×10^{10} cm/sec, the speed of light in free space. This was a surprise at the time, but today it is recognized that the electrostatic and electromagnetic units are related by the equation

$$\mu_0 \varepsilon_0 = \frac{1}{c^2} \, ,$$

where μ_0 is the permittivity of free space, ε_0 is the dielectric constant of free space, and c is the speed of light in free space. Just six years earlier Hippolyte Louis Fizeau (1819–1896) and Jean B. Léon Foucault (1819–1865) used a rotating mirror to determine the speed of light in air, arriving at the value 3×10^{10} cm/sec. (Foucault also measured the speed of light in different transparent media, and these results provided convincing evidence for the wave nature of light.)

Important as it was to have the electrical quantities related to mass, length, and time, it was even more important to find electrical standards that could be used in a practical manner. The first step in that direction was taken by Moritz von Jacobi of St. Petersburg in 1851. Jacobi suggested a practical unit of resistance, which found acceptance and became known as the Jacobi standard. The later acceptance of the ohm as the standard of resistance and its expression in absolute units is described in Section III.3.

For the forces between electric charges and currents, the entire scientific world used Coulomb's and Ampère's laws, both of which were "action-at-a-distance" force laws in the style of Newton's gravitational law. Though most people simply accepted that this was how nature behaved, some scientists sought a more comprehensible account of the actions of electric and magnetic forces. One of these was Michael Faraday.

Faraday continued to reflect upon how a magnetic field could produce an electric current, and vice versa. He often used metal filings to visualize a magnetic field, and this practice no doubt stimulated his view of the space between the interacting elements as being involved in the forces exerted. He argued that forces did not occur along the straight lines connecting any two elements, but rather along curved resultant "field lines." In 1838 he introduced the concept of electrostatic induction, and he saw this as an effect of the medium, as charges on the surface of a conductor became positioned at the end points of field lines normal to the surface. He also saw the medium as operating when dielectrics were used to increase the capacitance of condensers, and as a measure of this property he introduced the "specific inductive capacity" (which is now called the dielectric constant).

Faraday suffered from poor health, and in 1841 he left London and the Royal Institution for recuperation in Switzerland. When he returned four years later, he resumed his efforts to find some interaction between light and electric and magnetic fields.

Using the large magnet of the Royal Institution, Faraday discovered in November 1845 that the plane of polarization of a light beam was rotated as it passed through a strong magnetic field. The effect became known as the Faraday magnetooptic effect. Shortly thereafter, Faraday wrote in a letter to a friend that "the propagation of light might some day be accounted for by electromagnetic vibrations." How strange such an idea was at the time can be seen from the remark of John Tyndall (1820–1893), who described it as "one of the most singular speculations that ever emanated from a scientific man." (See Figure II.9.)

In his experiment with the light beam, Faraday had used a bar of so-called "heavy glass," that is, silicated borate. He discovered, to his amazement, that the bar oriented itself at right angles to the magnetic field lines. He knew that most materials aligned themselves with the field lines. Faraday then introduced the terms "paramagnetic," for the more common substances, and "diamagnetic," for those behaving as silicated borate did. This discovery of diamagnetism was one of his last experimental contributions.

Faraday's reflections on magnetic field lines had a great influence on James Clerk Maxwell, whose mathematical development of the concept of field is described in Section III.3. In 1858 Faraday, now 67 years old, retired to a place near Hampton Court in Surrey. His health was declining, and his memory began to fail rather rapidly. The man whose extensive discoveries did much to usher in the electrical age died peacefully on August 25, 1867.

II.4 JOSEPH HENRY AND ELECTRICAL TELEGRAPHY

In the last section we saw how Michael Faraday, who grew up in poor circumstances in London, had the good fortune to be exposed to science through books. On the other side of the Atlantic, in Albany, New York, a similarly talented boy, Joseph Henry (1797–1878), also grew up in poor circumstances and was also drawn to science by books. (See Figure II.10.) It was at age 16 that Henry, while convalescing from an illness, chanced to read a popular account of science, most likely George Gregory's *Lec-*

Figure II.9 Faraday enjoyed great success as a lecturer for the general public. This drawing (courtesy of the Burndy Library) shows Faraday lecturing at the Royal Institution in 1855.

Figure II.10 A portrait of Joseph Henry (courtesy of the Burndy Library).

tures on Experimental Philosophy, Astronomy and Chemistry. His interest thus aroused, Henry began taking courses, at the same time as he was working to earn money, at the Albany Academy.

It was not, however, until 1823, Henry's 26th year, that the young man was able to earn money in a scientific pursuit. In that year T. Romeyn Beck, principal of the Albany Academy, engaged Henry as assistant for a chemistry lecture series. The following year Henry himself delivered a lecture ("On the chemical and mechanical effects of steam") at the Albany Academy. In 1825 Henry worked as a surveyor for New York State, and in 1826 Henry, at the age of 29, was appointed professor of mathematics and natural philosophy at the Albany Academy. His inaugural address was a mature survey of the uses of calculus in civil engineering.

Partly through topographical and geological field work (which involved geomagnetic measurements) and partly through the writings of William Sturgeon (1783–1850), Henry became interested in magnetism. Sturgeon, inspired by Ampère's solenoid, had built an electromagnet by bending a bar of soft iron into a "U" shape, varnishing it, and loosely wrapping a coil of wire around it. Henry undertook a systematic investigation of electromagnets. He found that by coating the coil wire with insulating material, he could wrap the wire much more closely. In fact, he developed two classes of magnets: one, which he called "quantity magnets," used heavy wires carrying large currents at lower voltage; the other, which he called "intensity magnets," contained many turns of fine wire, requiring higher voltages but carrying smaller currents. (See Figure II.11.)

Joseph Henry's success in building electromagnets of great lifting power brought him international attention. In 1829 he exhibited an electromagnet that could lift 650 pounds, a record for that time. Two years later he delivered a still larger magnet for his friend, the publisher Benjamin Silliman (1779–1864). It was able to lift 2086 pounds, an incredible achievement at the time.

Joseph Henry was well aware of the limitations of chemical batteries, which of course he used for his electromagnets. Hence, he saw no great future for electric motors as a means of driving mechanical devices. He did, however, construct an electrical counterpart to the steam engine, namely, a device that used magnetic attraction and repulsion to produce reciprocating motion. This proved that electricity could generate mechanical motion, but was not promising as a practical device.

It is ironic that Henry was just then on the verge of discovering electromagnetic induction, Faraday's discovery that helped bring on the age of electricity. Henry had observed, in late 1830 or early 1831, that when he interrupted the current supply to his electromagnet a heavy spark occurred, which was,

Henry believed, "electricity produced from magnetism." This suggested to him that a change of magnetism was accompanied by an electric effect, which he identified as self-induction. He then wound an insulated wire around the iron armature with a galvanometer in the loop. When the electromagnet was activated, he observed a momentary deflection of the galvanometer. He also observed sparking if he left the galvanometer circuit slightly open when he interrupted the battery power to the main circuit.

On November 8, 1831 he received a letter from Parker Cleaveland (1780–1838), a professor at Bowdoin College, Maine, inquiring about the cost of a magnet similar to the one Henry had supplied to Benjamin Silliman of Yale College. Henry replied

> I have lately forged a large horseshoe weighing 101 lbs. which I intend fitting up for some contemplated *experiments on the identity of electricity and magnetism*. It will be much more powerful than any heretofore made being almost double the weight of the Yale College magnet (59 ½ lbs.)...

Figure II.11 A photograph of one of Joseph Henry's electromagnets in its test frame (courtesy of the Smithsonian Institution).

From this statement it appears that Henry intended to use the more powerful magnet to demonstrate electromagnetic induction. He wrote a paper entitled "The production of currents and sparks of electricity from magnetism" that was published in Silliman's journal (*The American Journal of Science*) in July 1832.

The late spring and early summer of 1832 were eventful for Henry. He sent the just-mentioned paper to Silliman, who in turn advised Henry of the announcement by the Scottish physicist James D. Forbes (1809–1868) of having obtained sparks from a "natural magnet." Henry accepted the offer of the chair of natural philosophy at the College of New Jersey (later Princeton University). And Henry learned of Faraday's publication in February on "Volta-electric induction and magneto-electric induction," which caused him regret at having been slow in publishing his own results.

For the next two or three years, Henry was completely absorbed with preparation of his lectures and demonstrations. He complained that "before I was called to the Chair of Natural Philosophy in this Institution, no lectures had been given on the subject nor experiments shown for many years." On October 22, 1833, Henry wrote in his notebook that while Faraday had demonstrated electromagnetic induction with the earth's magnetic field, a larger effect could be achieved using copper ribbon conductors specially wound.

Henry followed also the work of Carl Friedrich Gauss and Wilhelm Weber. In 1833 they used the newly discovered electromagnetic induction to produce current impulses for transmission over wires strung between the two observatories at the University of Göttingen. Though Gauss and Weber had only scientific interests in this achievement, they engaged C. A. Steinheil to make practical improvements in the method.

This was not the first attempt to use electricity for signaling. Already in 1820 Ampère recognized the possibility of using the effect of current upon magnetic needles to signal over wires, and Ampère even experimented with primitive coding. Others used the chemical action of electric current as an indication of a signal.

Henry was now located in Princeton, fairly close to Philadelphia, the home of the first American scientific society, the American Philosophical Society, established by Benjamin Franklin in

1743. Henry was pleased to be elected a member in 1835. In a letter on December 17, 1834, to Alexander Dallas Bache, a prominent member of that society, Henry asked, "Can you give me any information about the beautiful theory established by Ohm? Where is it to be found? I am anxious to see his papers." The fact that Henry had difficulty learning about Ohm's work, seven years after it was first presented, resulted partly from the geographic separation, but more from the slowness with which the scientific community accepted the findings of Ohm.

At Princeton Henry constructed the largest electromagnet yet; it was able to support a weight of 3600 pounds. And in 1836 he built a telegraph line across the campus, which he used to send signals from his home to the laboratory.

In February 1837 Henry left on an eight-month European tour. Because of his work on large magnets, he was generally received as a peer by the European scientists, but was deeply disturbed by the generally low opinion that Europeans, particularly the British, had of American science. Henry thereafter cultivated contacts with European scientists. For example, after presenting his work on electromagnetic induction (including self-induction) at a meeting of the American Philosophical Society on November 2, 1838, Henry sent copies of the subsequent publication to his European contacts.

There were at this time various attempts to use electromagnetic devices for signaling. Someone who saw commercial potential in a system of telegraphy was the British physician, William Fothergill Cooke (1806–1879). On a visit to Heidelberg to study anatomy, Cooke saw an experimental signaling system that used magnetic needles as indicators according to a primitive code. After he returned to London, Cooke tried building a system himself, but, lacking understanding of the phenomena, turned to Michael Faraday. Faraday referred him to Charles Wheatstone (1802–1875), professor at King's College in London, who was himself experimenting with telegraphy. Cooke and Wheatstone formed a partnership and applied at once for a patent. Issued on June 12, 1837, it was the world's first patent in telegraphy. But even Wheatstone had at that time limited comprehension of current-voltage relationships in networks. It was about this time that Joseph Henry, on his European tour, visited Wheatstone's laboratory and probably described his own telegraph.

Impatient to make a business of telegraphy, Cooke approached the London and Birmingham Railway Company. They agreed to a test installation of a telegraph line from Euston Station in London to Camden Town, and the first message was successfully sent on September 4, 1837. Cooke then approached the Great Western Railway, which agreed to a trial line from Paddington Station in London. This line, originally 13 miles in length, was of much improved structure and went into service in 1839. Although the provision of ground return and the use of a special code reduced the number of transmission wires, the cost of installation and operation militated against additional lines. Nevertheless, in 1845 the English Telegraph Company was organized, and it bought the relevant patents of Cooke and Wheatstone.

A contemporary of Faraday's and Henry's was Samuel Finley Breese Morse (1791–1872). All three became interested in electrical research, but they came from quite different backgrounds. Faraday started as apprentice bookbinder, Henry wanted to go into theater, and Morse became a successful artist. But all three were most strongly attracted by the mysterious manifestations of electricity.

Morse, whose father was an eminent Calvinist theologian, studied at Yale College. Upon graduation in 1811 Morse went to England to study the art of painting with Washington Allston. Returning to America in 1815, he specialized in portrait painting. He became one of the founders of the National Academy of Design in New York and served as its president from 1826 to 1845. In 1829 Morse returned to Paris to continue his art studies. It was on his return from Paris in 1832 that he learned from a fellow traveler about experiments in Heidelberg with electric telegraphy.

His interest thus aroused, Morse began electrical experimentation, though his position was that of professor of arts and designs at University of the City of New York. He was fortunate to gain the assistance of Leonard Gale, a faculty colleague who was familiar with Joseph Henry's work. It was, however, not until September 4, 1837, that Morse was able to demonstrate a telegraph system (see Figures II.12 and II.13): he transmitted a message 1700 feet using an electromagnet to move a pen into contact with paper.

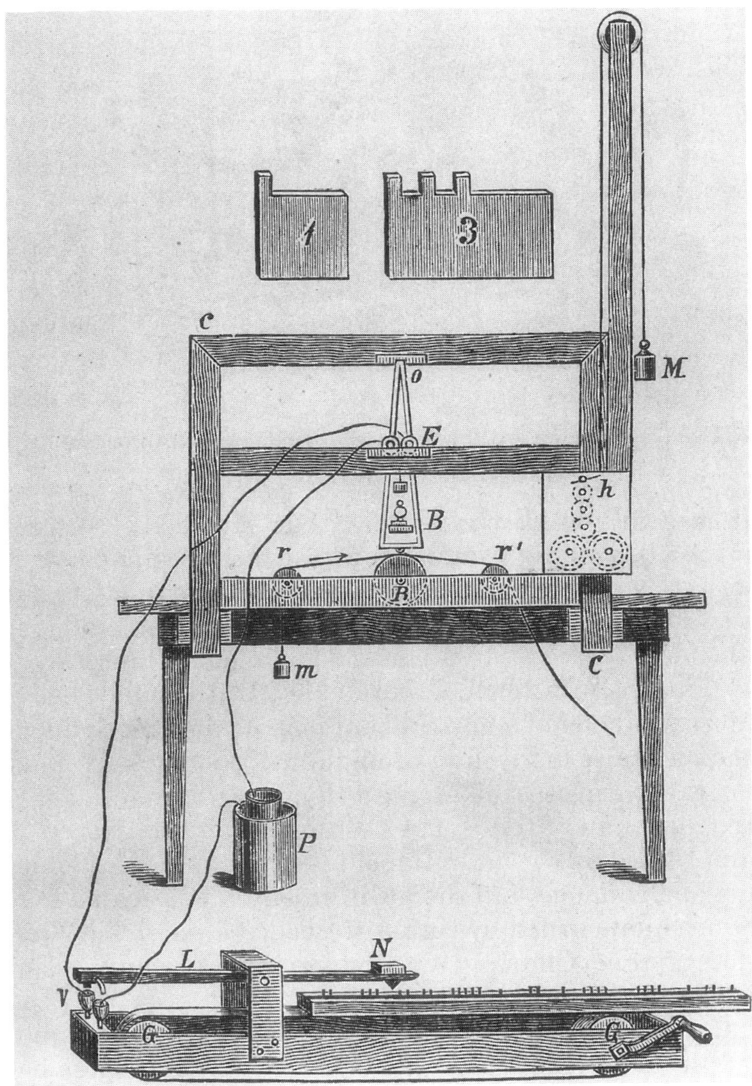

Figure II.12 A depiction of Morse's first telegraph. When one turned the crank at the bottom of the illustration, lever L moved up and down, closing and opening a circuit connected to the recording mechanism in the upper part of the illustration.

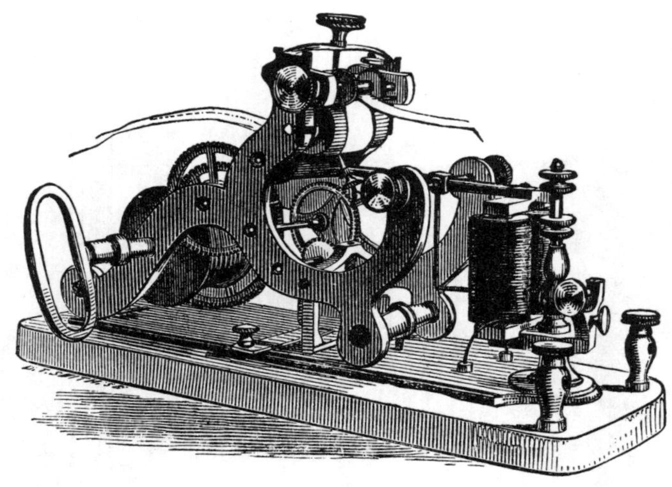

Figure II.13 An illustration of a Morse telegraph recorder, circa 1860.

It was at this demonstration that Morse met Alfred Vail (1807–1859), a young technician whose father owned iron works in Morristown, New Jersey. The elder Vail was ready to support Morse if he took his son as a partner. Morse agreed, and just four months later Morse and Vail gave a much more impressive demonstration in Morristown. Other demonstrations followed.

Morse learned of a government offer to aid the development of coastwise telegraphy, and he applied at once. It happened that one of the witnesses of Morse's demonstrations was Francis Smith, chairman of the House Committee on Commerce. Smith was impressed by the potential of telegraphy, but economic conditions had changed and the government offer became inactive. Morse was now urged by Smith to secure a patent in England, but found that Charles Wheatstone had already secured a patent there.

Later in 1838 Morse and Vail developed the telegraph code named after Morse. (With slight modifications it later became the International Code.) Vail added the sounder to their system, which facilitated the reading of the messages. Morse sought technical assistance from Joseph Henry, which he received both by correspondence and during Morse's visit to Henry at Princeton in May 1839.

In 1843 the government appropriated $30,000 to build a telegraph line from Baltimore to Washington, D.C. Morse built an overhead line (as suggested by Ezra Cornell) suspended from trees and poles along the Baltimore and Ohio Railroad. The line was completed and the first message sent on May 24, 1844— "What hath God wrought?" The telegraph service was an immediate success. Initially, the post office took charge, but by 1845 that support stopped, and Morse with his partners formed the Magnetic Telegraph Company, a private stock company.

As time went on, more and more companies, paying royalties to Morse and his partners, went into the telegraph business. In 1851 Hiram Sibley (1807–1888), who had followed Morse's activities since 1844, organized the New York and Mississippi Valley Printing Company, which bought up 11 of the smaller companies, and in 1856 Sibley formed the still larger Western Union Telegraph Company. By then telegraph lines were extending all over the United States, and in 1861 a transcontinental line was completed. By 1876, 250,000 miles of telegraph wires were in use.

The success of telegraphic communication, first demonstrated in Great Britain and then more convincingly in the United States, could not but invite additional enterprise. In Germany, Werner von Siemens (1816–1892) had just finished his required military service in 1845 when he learned of Cooke and Wheatstone's telegraph and saw a model telegraph in Berlin. He became interested and, having already received a technical education, was able to devise several improvements. Werner Siemens then formed a partnership with his brother Georg and with the mechanic Johann Georg Halske to establish the Siemens & Halske Company. This company soon became a main supplier of telegraph equipment in central Europe.

It is not surprising that the development of the electric telegraph gave rise to contested claims of priority. Morse, whose U.S. patent was issued in 1842, became involved in litigation with his former partners. He quarreled with Joseph Henry and finally broke off this relationship. Henry felt that he had contributed essential ideas, whereas Morse believed that it was only the final product, which he had constructed, that counted. The case eventually went to the Supreme Court, which ruled in favor of Morse.

The dispute between Henry and Morse concerned one aspect of the relationship between science and technology, namely, to what degree a technological advance is straightforwardly an application of scientific understanding. Henry's view of the telegraph was expressed in a letter to Morse in 1842, before their falling out:

> ...in the minds of many, the electro-magnetic telegraph is associated with the many chimerical projects constantly brought before the public and particularly with the schemes so popular a year or two ago for the application of electricity as moving power—all schemes for this purpose, I have from the first asserted, are premature and formed without proper scientific knowledge. The case is different in regard to the electro-magnetic telegraph. The science is now fully ripe for such an application of its principles, and I have not the least doubt...of the perfect success of the invention.

Henry and Morse illustrate also two personality types, the scientist and the engineer: Henry took the observation of sparks as a key to greater understanding, while Morse saw in sparks the possibility of signaling. (The case of the Atlantic cable, considered shortly, is more clearly an example of "scientific engineering," that is, engineering dependent upon scientific knowledge.)

Henry continued his exploration of electrical phenomena; in 1842, for example, he observed the oscillatory discharge of the Leyden jar. But other activities soon claimed most of Henry's time. In 1846 he was selected to head the newly created Smithsonian Institution. One of his first large projects at the Smithsonian was to organize a corps of volunteers to report weather conditions using the rapidly expanding telegraph network.

During his 1837 tour of Europe, Henry had attended a meeting of the British Association for the Advancement of Science, which had been founded in 1831. Upon his return to the United States, he conferred with others about establishing an American counterpart, but nothing came of these discussions. After he became secretary of the Smithsonian Institution, however, he felt more strongly that scientific efforts and interchange should be promoted, and he joined with others to establish the American

Association for the Advancement of Science in 1848. Henry, indeed, served as AAAS president in 1849.

Henry served also as science advisor to government. From 1852 until his death, he was a member of the Federal Lighthouse Board, and in that capacity he sought to increase the efficiency of the beacons used for coastal navigation. And during the Civil War President Lincoln turned to Henry for scientific advice. Henry, working with his friend Alexander Dallas Bache, institutionalized this role by forming the National Academy of Sciences, which was chartered by Congress in 1863.

We turn now to the subject of submarine telegraphy. In 1842, in one of his early demonstrations, Morse laid copper wire across the New York harbor and sent messages through it. The wire was covered with gutta-percha, a resinous material that had recently been recognized as an excellent insulator. And in the following year Morse wrote in a letter to the secretary of the Treasury that he foresaw telegraphic communication across the Atlantic.

The first use of a submarine cable for commercial service was laid across the English Channel in 1850; after a short period of operation it was broken by accident. On the basis of the initial success, however, several other submarine telegraph connections were established, such as from Corsica to La Specia, all with cables manufactured in England.

In North America, telegraph land lines soon extended into Maine, and it was proposed to go on into Newfoundland. Because of financial difficulties, however, a shorter submarine cable from New Brunswick to Prince Edward Island was laid in November 1852. This work was directed by Frederic N. Gisborne, who had organized the Newfoundland Electric Telegraph Company. In order to obtain more financial aid, Gisborne went to New York and succeeded in arousing the interest of Cyrus Field (1819–1892).

Field was especially taken by the idea of a transatlantic cable. He joined with some friends—among them Peter Cooper (1791–1883), who a few years later founded the engineering school Cooper Union—to form the New York, Newfoundland, and London Electric Telegraph Company. It was designed as an international company, to operate in the United States, Canada, and Great Britain. Peter Cooper was elected president on May 6,

1854, and capital was secured by stock subscriptions. The first job was to complete the telegraph land line across Newfoundland over very rough territory and then lay a cable from Cape Ray across the Gulf of St. Lawrence to the island of Cape Breton and thence to Nova Scotia. There were some delays, but the line was completed in 1856.

Though Morse, on the basis of tests on a short underwater cable, concluded that signals could be propagated at satisfactory speed underwater, doubts remained. In 1855 William Thomson, later Lord Kelvin (1824–1907), studied the problem mathematically. Thomson made use of the method devised by Joseph Fourier to solve the one-dimensional heat diffusion problem. Thomson's solution, which disregarded the magnetic field as well as the leakage current in the dielectric, gave the so-called K-R law (K being the total capacitance and R the total resistance of the cable). According to this law, the buildup time of the signal would increase with the square of the distance, which implied that a very long cable would be very inefficient. There was the additional problem that electrical resistance in a 2000-mile copper line would reduce the signal dramatically. To solve this problem, Thomson devised an extremely sensitive mirror galvanometer (which would be observed by one person while a second person kept a record of the reported movements).

Building a transatlantic cable was a tremendous enterprise, requiring great financial resources and technical advances in cable manufacture; in machinery for storing, transporting, and laying out cable (which had to be done in waters 2 miles deep); and in transmission and detection apparatus. In 1856 when the English government agreed to join in the financing, the name of the organization was changed to the Atlantic Telegraph Company. At the same time William Thomson was added to the Board of Directors.

The first expedition started from Valentia Harbour in England on August 5, 1857, but experienced cable failure after five days. The following year, another attempt was made. The cable was successfully laid across the Atlantic to Trinity Bay, Newfoundland, where it was connected to land lines. On August 16, 1858, in New York, Cyrus Field first tested the cable. Among the first messages was an exchange of greetings between Queen Victoria and President James Buchanan.

The signal received was, however, extremely weak. Moreover, Thomson's K-R law was confirmed: it took 16 hours to transmit the queen's message of 98 words, and the president's reply of 149 words took more than 10 hours. After just a month of operation, the cable failed permanently.

Seven years passed before the next attempt was made. A committee of eminent scientists (including William Thomson) and telegraph engineers (including Charles Wheatstone) submitted a report on July 13, 1863, with their conclusion that a well-built cable could now be laid that would function over many years. Encouraged by this report, and despite the ongoing Civil War, Cyrus Field proceeded vigorously planning a new cable.

Field's new company, the Telegraph Construction and Maintenance Company, used entirely new machinery and ships. In July 1865 it began laying the new cable, but again a fault developed. After a positive report by experts, Field obtained additional financing and formed a new company, the Anglo-American Telegraph Company. This time there was complete success, the cable having been laid by the steamship *The Great Eastern* in July 1866. (See Figure II.14.)

Submarine telegraphy expanded rapidly. Already in 1869 a French transatlantic cable was laid, and soon cables connected Japan to the Asian mainland and New Zealand to Australia. By 1880, 90,000 miles of submarine cable were in use.

Even before the widespread use of submarine cables, international telegraphy was an important enterprise. In 1865 the International Telegraph Union (ITU) was formed. With headquarters in Geneva, the ITU played a large role in bringing about international agreements on telegraph practice. Telegraphy continued its advance as theoretical studies, inventions of new devices (such as Thomson's siphon recorder, patented in 1867), new practices (such as duplexing messages using bridge circuits), and the establishment of schools for operators all led to improved service.

II.5 TELEPHONY

It was not long after the establishment of electrical communication by telegraphy that people speculated about transmitting

Figure II.14 An illustration of loading the cable, which was more than 2000 miles long, into *The Great Eastern* (courtesy of the Burndy Library).

human speech electrically. Telegraphy, of course, used simple dot-dash combinations requiring only the making and breaking of a current, either from a battery or a DC generator. Speech, on the other hand, consists of mechanical vibration of the air over a substantial frequency range, as Hermann von Helmholtz had demonstrated in the 1860s. People sought ways to convert the mechanical vibrations to variations in electrical current, which could then be sent through a wire. As early as 1861 a German school teacher, Philip Reis, succeeded in transmitting musical tones electrically; despite arousing considerable interest, this invention was not developed further.

Something that was pursued with vigor in many countries was multiplex telegraphy, that is, sending two or more messages over the same line simultaneously. In the United States, Elisha Gray (1835–1901) directed his inventive talents to this endeavor. By 1867 he had several patents on multiplex telegraphy and had set up a small company in Cleveland to exploit them. Operations were moved to Chicago when Western Union, which had been a customer, acquired part ownership in 1872. After further expansion, Gray's company changed its name to Western Electric Manufacturing Company.

In 1874 Gray resigned as superintendent of the growing enterprise in order to give his full attention to his experiments with "vibrating currents." He had found that by placing a poor conductor between two metal conductors (serving as what would later be called a microphone), he could transmit musical tones electrically. After he succeeded in transmitting chords, that is, composite tones, he gave a demonstration of this "musical telegraphy" at the Smithsonian Institution in June 1874. This was witnessed by Joseph Henry, who was much impressed.

Though Gray stressed the scientific aspects of his discovery, he had an eye to commercial exploitation. He thought his invention might make the transmission of speech possible, but he had no doubt that the most significant application was multiplex telegraphy. Telegraphy was, after all, an established business, and the financial remuneration for improvements could be immediate and substantial.

At the same time as Elisha Gray, another person, from an entirely different background, was also making progress on multiplex telegraphy. Alexander Graham Bell (1847–1922) was born in Edinburgh, Scotland. His father, Alexander Melville Bell, who was professor of elocution at the University of Edinburgh, had developed a system of "visible speech." This was a new and highly successful method of teaching the deaf to speak, and it brought Alexander Melville Bell international recognition. He encouraged his son's interest in physiology, especially the mechanism of human speech, and in the use of tuning forks. Young Bell's interests made Alexander Ellis, well-known phoneticist, refer him to Helmholtz's treatise on the human perception of tones, which the young Bell immediately acquired.

In 1870 the Bell family emigrated to Canada, settling in Brantford, Ontario, and the following spring the father helped his son gain a teaching position at the School for Deaf Mutes in Boston. The young Bell, applying, of course, his father's methodology, was successful both as a teacher of the deaf and as a public lecturer. In 1873 he was appointed professor of vocal physiology and elocution at Boston University.

At about this time Bell learned of Elisha Gray's work on multiplexing, as well as his musical telegraphy. With his experience with tuning forks, he could at once see the possibility of using selected frequencies for transmission of telegraphic messages. However, he was, like Gray, unable to achieve an effective separation of the frequencies at the receiving terminal. Recognizing that Gray was a formidable rival in the effort to achieve multiplexing using different frequencies, Bell resolved to concentrate entirely on speech transmission. He also began to be more secretive about his work.

Bell accepted the job of private tutor to the deaf son of Thomas Sanders, a leather merchant, and moved to the Sanders's home in the seaport community of Salem, north of Boston. In Salem he met a lawyer, Gardiner Greene Hubbard, whose deaf daughter Bell married four years later, on July 11, 1877. In February 1874, Sanders and Hubbard concluded a contract with Bell to support his developments financially and to share in the eventual income.

Bell consulted with the inventor of the fire alarm telegraph, Moses G. Farmer, and engaged the shop of Charles Williams in Boston to build a model of his design for a telephone. Thomas A. Watson, apprentice in Williams's workshop, became Bell's assistant. Bell's design used slightly magnetized metal reeds both in a transmitter and in a receiver.

By chance, on June 2, 1875, Watson, while repairing the model in one room, transmitted sounds of the reeds to the neighboring room where Bell was present. Together they reconstituted the occurrence, improved the arrangement, and added small horns as speakers (see Figure II.15). On February 14, 1876, Hubbard, representing Bell, applied for a patent. The application described the rather crude model, but mentioned the possibility of varying the circuit resistance and thereby the current by immersion of a metal rod in liquid. A few hours later on the same day, Elisha Gray's patent attorney submitted a caveat (which is priority declaration, not a patent application).

With further improvement, Bell could demonstrate clear transmission of speech. The most celebrated demonstration occurred at the U.S. Centennial Exhibit in Philadelphia on June 15, 1876, which was witnessed by J. A. Fleming of England and Don Pedro, emperor of Brazil (who reportedly marveled that the new invention was able to speak Portuguese). William Thomson praised Bell's transmission of speech as "perhaps the greatest marvel hitherto achieved by the electric telegraph."

Just over a year separated the first demonstrations of the telephone and commercialization of the invention. On October 9, 1876, Bell and Watson held a two-way conversation over telegraph wires from Boston to Cambridge. Less than two months later they conversed across the 143 miles separating Boston and Conway, New Hampshire. The first commercial offer came in April 1877, and on July 9 Hubbard incorporated the Bell Telephone Company. Bell (see Figure II.16) served as technical advisor, staying on the board of directors until 1879.

An important advance in telephone technology was an improved microphone. In 1877 Emile Berliner invented an effective microphone that depended upon variable contact. In the same year Thomas Edison invented a carbon transmitter, but the Bell Telephone Company used a carbon transmitter invented by Francis Blake.

In 1876 Hubbard offered to sell the rights to Bell's telephone patent to Western Union for $100,000. The offer was rejected. But with the success of Bell's demonstrations of his telephone, Western Union organized the American Speaking Telephone

Figure II.15 A photograph of Alexander Graham Bell's second telephone (the *gallows telephone* of 1875) (courtesy of the Smithsonian Institution).

Figure II.16 A photograph of Alexander Graham Bell.

Company as a competitive organization. In 1878 the Bell interests brought suit, claiming patent infringement, and on November 10, 1879, a settlement was reached. Western Union acknowledged the validity of Bell's patent and agreed to withdraw from the telephone business, and the Bell Company agreed not to compete in the telegraph business.

In 1880 the Bell Company changed its name to American Bell Telephone Company, and two years later it purchased

Western Electric Manufacturing Company of Chicago to manufacture its equipment. In 1885 a separate company, American Telephone and Telegraph (AT&T), was formed to construct and operate intercity lines. On December 30, 1899, all assets of American Bell were transferred to AT&T, which thus became the parent company. By this time the number of telephones in the United States had increased to more than 1.5 million.

In England, France, and Germany telephone systems developed, though not so rapidly. In each of these countries it was a government agency that provided telephone service, so the prospect of large profits did not provide the same inducement to development that they did in the United States.

II.6　The First Central Power Stations

The first electric generators were, as we saw in Section II.3, simply one or more wire coils that rotated past a bar magnet, inducing a feeble alternating current. Since at that time voltaic piles were the main source of electricity, it was direct current that was seen as useful. The first generators were, therefore, equipped with commutators, which gave a unidirectional, but pulsating, current. Stimulated by the need for sources of current in telegraphy, William Ritchie (1790–1837) and Joseph Saxton (1799–1873), both working in London, made improvements to the electric generator in the 1830s. Another person who contributed to the improvement of generators in that decade was Edward Montague Clarke (ca. 1804–1846), an electrical instrument maker, who had developed a magnetoelectric machine by 1836 that promised "uniform" current supply. Clarke published a report on his machine in *Transactions and Proceedings of the London Electrical Society*, which was one of the first publications devoted to electrical science and technology. It must be borne in mind that at that time theoretical understanding, including the concept of the magnetic circuit, and practical techniques, including an efficient means of insulating the wires in a coil, were frequently lacking.

Progress continued in the 1840s. Moritz Hermann de Jacobi (1801–1875), born in Germany but working in Russia, made improvements in the armature structure of the generator in 1841.

The following year James Prescott Joule (1818–1889), who had observed the heating of current-carrying wires, established the fact that the energy lost in a resistive circuit is the product of the resistance and the square of the current. Moreover, he established the equivalence of electrical energy and heat energy. His work was an important element in the recognition of the law of conservation of energy, which came about through the work of Julius Robert von Mayer, Justus von Leibig, Hermann von Helmholtz, and others in the 1840s and 1850s.

A notable milestone in the improvement of electric generators came in 1845 when Charles Wheatstone (1802–1875), who was working in telegraphy, replaced the permanent magnets with electromagnets. In 1856 Werner von Siemens (1816–1892) built a generator with the armature in the form of a double T, better to accommodate the copper coils. The following year, he recognized the possibility of self-excitation, namely, feeding the generated current directly into the electromagnets. (Before this, the electromagnets had been powered by their own source of current.) In 1860 Antonio Pacinotti (1841–1912) placed the armature windings into slots, and this gradually became the favored design. A decade later Zenobe Theophil Gramme (1826–1901), working for the Alliance Company in Paris, designed a DC generator similar to Pacinotti's, and Gramme's generator (see Figure II.17), which could also function effectively as a motor, became fairly standard.

By this time the variety of units for measurement of electrical quantities had become a concern. It is true that the work of Gauss and Weber gave relationships between the cgs metric system and absolute measurements of the electrical quantities (as discussed in Section II.3), but there was still a pressing need for practical units of measure and for reference standards. In response, in 1861 the British Association for the Advancement of Science appointed, at the urging of William Thomson, a committee to formulate proposals. The committee's report, delivered in 1863, introduced the volt as the unit for electromotive force, defining it as 10^9 cgs units, and the ohm as the unit for resistance. Formal international acceptance on these two units came in 1881 at the First International Congress of Electricians in Paris, where the ampere (for current), the coulomb (for charge), and the farad (for capacitance) were also accepted.

Figure II.17 An illustration of a Gramme generator.

As DC generators became more easily available, the application of electricity to medicine increased under the leadership of Guillaume Benjamin Armand Duchenne (1806–1875). Duchenne showed how to use induction coils to achieve what he called "localized electrization," which could be of diagnostic and therapeutic value. The study of "animal electricity," which became known as electrophysiology, started with the work of Luigi Galvani (1737–1798), but made great progress at the hands of Emil Du Bois-Reymond (1818–1896), professor of physiology at the University of Berlin.

The availability of electric generators stimulated activity in another realm, the production and study of electrical discharges. By using induction coils one could obtain a high voltage in the secondary winding, enough to cause an electric discharge between two electrodes. In 1838 Faraday had observed a glow near the cathode as well as near the anode with a dark space in between, called Faraday dark space. A decisive advance was the invention in 1855 of a highly effective vacuum pump (the mercury air pump) by Johann Heinrich Wilhelm Geissler (1814–1879) in Bonn. One could obtain various luminous effects with electrodes placed inside tubes evacuated to very low pressures, which came to be called Geissler tubes. A number of people studied discharges under differing conditions, and this led to the detection of "cathode rays," so called by Eugene Goldstein (1850–1930) of the Berlin Observatory. Perhaps the most illustrious of these investigators was William Crookes (1832–1919), who in the early 1880s made a series of influential studies of electrical discharges. It was with a modified Crookes tube that Wilhelm Conrad Röntgen (1845–1923) discovered the x-rays in 1895. (In most languages, x-rays are called Röentgen rays.)

One of the most significant applications of the new electric generators was for illumination. Made practical for the first time by the Gramme generator was carbon arc lighting, which had been demonstrated by Humphry Davy in 1802 using a large battery. Arc light searchlights were used by the French in the Franco-Prussian War of 1870 and 1871. In 1876 Paul Jablochkoff (1847–1894) invented the Jablochkoff candle, which used two parallel carbon rods insulated from each other by kaolin. Gramme used alternating current with these lights to keep the two rods burning down at the same rate. Soon lighthouses were equipped with arc lights, and the public places in Paris (see Figure II.18), London, and Berlin became brilliantly lit at night. In the United States, Charles Francis Brush (1849–1929) developed similarly efficient arc lamps, powered with the Brush dynamo, and Brush's arc lamps lighted the public square in Cleveland (Brush's hometown) in 1879 and part of New York City in 1880.

The Brush Company became a leading producer of dynamos, but in 1889 it was acquired by the Thomson-Houston Electric Company. The latter company began as the American Electric Company, founded by Elihu Thomson (1853–1910) and Edwin J. Houston (1847–1914) in 1880. In the late 1870s Thomson and Houston collaborated in designing a dynamo and an arc lamp, and then decided to form a company to produce the dynamo and entire arc-lighting systems. After a slow start, the company was quite successful. Three years after acquiring the Brush Compa-

Figure II.18 The Avenue de l'Opera in Paris illuminated by Jablochkoff arc lamps.

ny, Thomson-Houston joined with the Edison companies to form General Electric.

It was to provide arc lighting that the first central power station in the United States was built. This occurred in San Francisco in 1879, when the newly formed California Electric Light Company put a small power plant—with Brush generators—into service.

The proliferation of central power stations did not, however, occur until incandescent lighting had become practical. In England Joseph Wilson Swan (1828–1914) experimented with incandescent lamps, and on December 28, 1878, he demonstrated a practical lamp that contained a carbon filament. He improved the lamp, obtained a patent in 1880, and formed the Swan Electric Lighting Company.

In the United States it was Thomas Alva Edison (1847–1931) who did the most to make incandescent lighting practical. Edison was working as a telegrapher in the mid-1860s when he learned of Faraday's work and began experimentation himself. His first successful invention was a stock ticker, for which he submitted a patent application in 1869. He made improvements in multiplex telegraphy and invented both the carbon transmitter (or microphone) and the phonograph in 1877.

In 1878 Edison founded the Edison Electric Light Company, and, working with his assistant Francis Upton, sought filaments with high resistance. After many attempts with carbonized materials, they were successful with carbonized cotton thread on October 21, 1879. Edison made a large public demonstration of his lamp on December 31, 1879, and obtained a patent on the device on January 17, 1880. He also applied for a patent in Great Britain, and this led to much litigation that was ended with the creation of the Ediswan Company, combining the English Edison and the Swan companies.

Edison realized that a distribution network for electric power generated at a central station would create a tremendous market for incandescent bulbs and for electricity in homes, businesses, and factories. At the Paris International Electrical Exhibition in the fall of 1881, Edison presented a large generator of the Gramme type with 50-kW output powering a lavish display of incandescent lighting. A tragic incident that December underscored the advantage of electric lighting over gas lighting: the

Ring Theater in Vienna burned to the ground with great loss of life.

Edison now had no problem securing financial support, and on January 12, 1882, he put into operation a central power station at Holborn Viaduct in London; it furnished power for 3000 incandescent lamps. Shortly thereafter, on September 4, 1882, Edison's power station on Pearl Street in New York City (see Figure II.19) went into operation; its six jumbo generators furnished DC power for 7200 lamps at 110 volts (considered safe for public use). In the same month an Edison hydroelectric station at Appleton, Wisconsin, went into operation.

By late 1887 there were 121 Edison central stations functioning in the United States; they all distributed DC power at 110 volts and collectively powered more than 300,000 lamps. The tremendous success of the Edison Electric Light Company attracted competitors. The largest was the United States Electric Lighting Company formed by Edward Weston and Hiram Maxim; in 1888 it was taken over by the Westinghouse Corporation.

The success of the International Electrical Exhibition at Paris in 1881 led to a second international exhibition in Munich in 1882 and to a third in Vienna in 1883. The U.S. Congress provided funding for the United States Electrical Commission to host the fourth international exhibition in Philadelphia in 1884 (see Figure II.20). This exhibition stimulated the organization of a professional society akin to the British Institution of Electrical Engineers (IEE). [The IEE had been founded in 1871 as the Society of Telegraph Engineers, with Charles William Siemens (1823–1883) as first president, and the French Société des Électriciens had been founded in 1883.] Nathaniel S. Keith of New York was the instigator. A preliminary meeting on April 15 was followed by one on May 13, 1884, when the "American Institute of Electrical Engineers" was chosen as the name of the new organization and Norvin Green, president of Western Union, as first president.

The International Electrical Exhibition was held in Philadelphia from September 2 to October 11. The United States Electrical Commission, with Henry Augustus Rowland (1848–1901) of Johns Hopkins University as president, organized professional activities during the exhibit. Most significant was the confer-

Figure II.19 A drawing of the dynamo room of the Edison power station on Pearl Street in New York City.

Figure II.20 A poster prepared for the 1884 International Electrical Exhibition in Philadelphia (courtesy of the Franklin Institute).

ence, at which William Thomson gave the first address. Thomson stressed the need for international standards in the new field of electricity and magnetism. In the final days of the exhibition the American Institute of Electrical Engineers (AIEE) held its first membership meeting.

The most lavish display at the exhibition was Edison's. His generators and lights attracted the most attention. Also part of the Edison exhibit, however, was a three-electrode lamp. The previous year Edison had first demonstrated what came to be called the Edison effect: current would flow from a heated filament to a test wire if the test wire was connected to the positive terminal of a battery, but not if the wire was connected to the negative terminal. Though the device aroused considerable interest—at the October 7 meeting of the AIEE, Edwin T. Houston made a report on it—no one saw important applications. It is today seen as the forerunner of the electron tube.

The availability of powerful dynamos stimulated work on electrically powered public transit (see Figure II.21). The first public tramway system was developed by Werner von Siemens for the Berlin Industrial Exhibition in 1879. It was pulled by a small electric locomotive, driven by a DC motor supplied with power at 180 volts through a third rail, later replaced by overhead wire. For several years this line operated as a commercial tramway in the district of Lichterfelde; in 1881 it was extended to a distance of one and a half miles.

Its success stimulated developments in the United States. The first commercially successful trolley system was that completed in 1888 by Frank J. Sprague (1857–1934) for the city of Richmond, Virginia. Sprague had developed a motor suitable for electric traction, and he used an overhead wire (with return current through the metallic track) to supply power. Sprague was especially innovative in finding a satisfactory way of mounting the motors and transferring power to the axles. Sprague's chief rival, Charles Van Depoele (1846–1892), solved the problem of sparking of the brushes on the commutator with load changes: he replaced the copper brushes with carbon brushes. In a flood of development, American cities rushed to build electric trolley systems; by 1890 180 systems were in operation or under construction.

II.7 ALTERNATING CURRENT

By the 1890s, electrical communication by telegraphy and telephony had progressed to the establishment of national and in-

Figure II.21 One of the early applications of electric motors was in elevators, as seen here in a turn-of-the-century Otis advertisement (courtesy of United Technologies Corporation).

ternational networks, and many cities had direct-current generating and distribution systems. Nevertheless, by modern standards engineering design was in a rather primitive state, and as a rule electrical engineers made little use of mathematical theory in their work. This situation changed with the rapidly expanding use of alternating current in the 1890s and thereafter.

One should remember that it was only some 90 years earlier that Volta, with his discovery of a battery, showed how to obtain a continuous current and thus opened an immense new realm of phenomena. Electrochemical processes, the magnetic effect of currents, the relationship between electromotive force and current, and electromagnetic induction were, as we have seen, all investigated in the following decades. With the development of the telegraph and the telephone, both powered by batteries, and even with the building of direct-current distribution systems (with generators equipped with commutators to produce unidirectional current), it was possible to do satisfactory engineering without much use of mathematical theory.

To understand the electromagnetic characteristics of varying currents, however, required mathematics. Differential and integral calculus, independently invented by Isaac Newton (1642–1727) and Gottfried Wilhelm Leibniz (1646–1716), were essential. But in the nineteenth century only a few people attempted to apply such mathematical tools to the phenomena of electromagnetism.

One of these people was Hermann von Helmholtz. In 1851 he analyzed the so-called "extra" current, the current flowing in a closed circuit with inductance L and resistance R if the applied voltage V is removed. He found that the current would decay exponentially as the magnetically stored energy is dissipated:

$$i = \frac{V}{R}\, e^{-\frac{R}{L}t}.$$

William Thomson extended this analysis in 1853 by considering the transient response of a circuit with resistance R, inductance L, and capacitance C in series with constant voltage V. He found that the behavior of this circuit was described by a differential equation:

$$V = Ri + L\frac{di}{dt} + \frac{1}{C}\int i\ dt.$$

In addition to the exponentially monotonic decay Helmholtz had described, Thomson found that for the condition

$$R < 2\sqrt{\frac{L}{C}}$$

the circuit exhibits an exponentially decaying oscillatory response, where the frequency is given by

$$f = \frac{1}{2\pi}\sqrt{\frac{1}{LC} - \frac{R^2}{4L^2}}.$$

Under the ideal condition of $R = 0$, this becomes the natural frequency of the lossless double-energy circuit

$$f = \frac{1}{2\pi}\sqrt{\frac{1}{LC}}.$$

In 1857 Gustav Robert Kirchhoff treated the propagation of electric waves along a wire, and for the case of submarine telegraphy he included inductance (which William Thomson had left out of his treatment). Kirchhoff obtained an approximate value for the velocity of propagation, but his paper attracted almost no attention at that time. It nevertheless indicated what a person with appreciable mathematical skills could accomplish. (Another example is Kirchhoff's 1868 mathematical treatment of the propagation of sound in a narrow tube, which took into account viscosity as well as the effect of the walls; this was an important result that John William Strutt, Lord Rayleigh, drew upon for his 1877 book *The Theory of Sound*.)

The mathematical genius of James Clerk Maxwell illuminated some difficult problems involving alternating-current circuits. In 1865 he considered the situation of two single-energy, magnetically coupled circuits. For these he computed the steady-state response for a sinusoidal voltage applied to the primary circuit. He found that the magnetic coupling reduces the effective inductance and increases the effective resistance in the primary circuit.

Maxwell then calculated the response of a double-energy circuit to an applied sinusoidal voltage. He arrived at the equation

$$i(t) = \frac{V_m}{R\Delta}\left[\sin\omega t + \frac{1}{R}\left(\omega L - \frac{1}{\omega C}\right)\cos\omega t\right],$$

where V_m is the maximum voltage, ω is the angular frequency (that is, 2π times the frequency in cycles per second), and

$$\Delta = 1 - \frac{1}{R}\left(\omega L - \frac{1}{\omega C}\right)^2.$$

If the applied frequency coincides with the natural frequency of the circuit,

$$\frac{1}{2\pi}\sqrt{\frac{1}{LC}},$$

there is resonance, which is the largest possible current response. Maxwell confirmed this experimentally in 1868.

At about this time interest turned to certain geometrical relationships rediscovered by Jean-Robert Argand in 1806. He had argued that complex numbers (that is, numbers of the form $a + bi$, where a and b are real numbers and i satisfies the relationship $i^2 = -1$) may be effectively represented as line segments emanating from the origin of a cartesian coordinate system. Since any complex number can be expressed in the form $r(\cos \alpha + i \sin \alpha)$, r can be regarded as the length of the corresponding line segment and α as the angle between the positive x-axis and the segment. Multiplication of a complex number by i is then simply counterclockwise rotation of the corresponding line segment by $90°$. Indeed, multiplication of any two complex numbers is easily viewed geometrically: since

$$r(\cos\alpha + i\sin\alpha) \times s(\cos\beta + i\sin\beta)$$

$$= rs\left[\cos(\alpha + \beta) + i\sin(\alpha + \beta)\right],$$

the length of the product segment is the product of the lengths of the multiplier segments and the angle of the product segment is the sum of the angles of the multiplier segments. Since $e^{i\theta} = \cos\theta + i \sin\theta$, this result can be expressed more compactly as

$$re^{i\alpha} \times se^{i\beta} = rs\, e^{i(\alpha+\beta)}.$$

De Moivre's theorem,

$$[r(\cos\theta + i\sin\theta)]^n = r^n(\cos n\theta + i\sin n\theta),$$

becomes evident with this representation:

$$(re^{i\theta})^n = r^n e^{in\theta}.$$

If we now express an alternating current of amplitude I and sinusoidal variation in complex form by $i(t) = Ie^{j\omega t}$, we see that, as a function of time, the segment representing $i(t)$ rotates with frequency $f = \omega/2\pi$ in the counterclockwise sense. (We see also why in engineering texts the imaginary unit is designated j rather than i.) At a particular moment, $i(t)$ might have the position OP indicated in Figure II.22. The instantaneous current is given, at a given time, by the abscissa, which is to say the real part of the complex quantity $i(t)$.

Phase relations are perspicuously displayed in such diagrams. Consider the Argand representation of two currents, sinusoidally varying with the same frequency and differing in phase by the angle ψ (see Figure II.23).

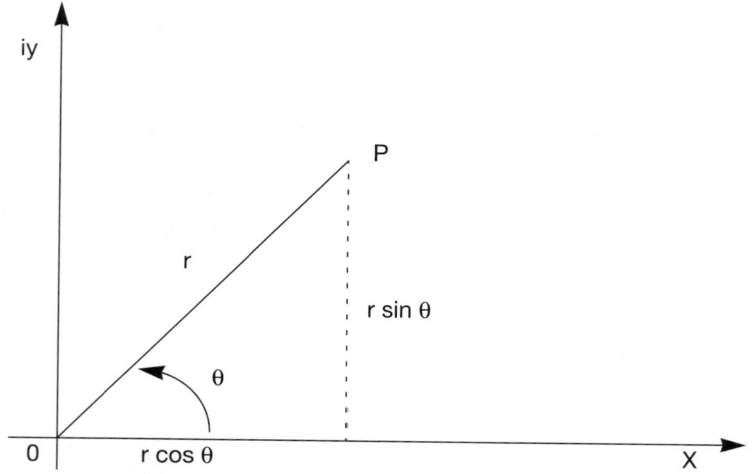

Figure II.22 The Argand representation of a sinusoidally varying current.

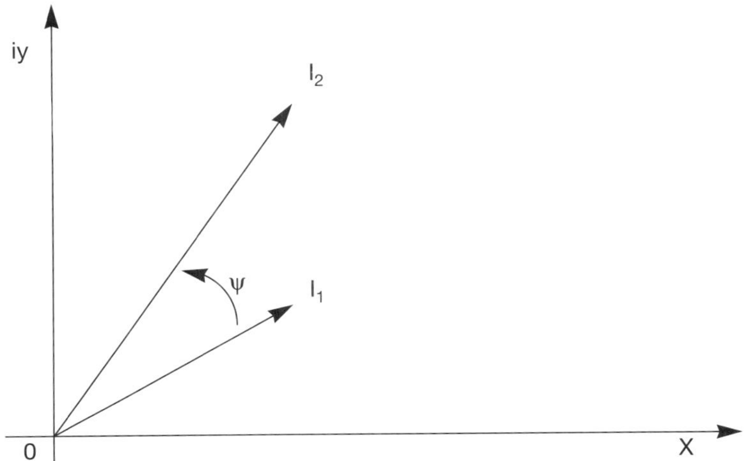

Figure II.23 The Argand representation of two currents, sinusoidally varying with the same frequency and differing in phase by the angle ψ.

This graphical representation has proved to be an effective tool for understanding complex circuits. Indeed, it was adopted as the first tangible way to portray the voltage drops for the elements of a circuit.

In the late nineteenth century people had great difficulty predicting the behavior of alternating current in complex circuits, such as coupled networks. In 1888 Oliver Lodge (1851–1940) commented, "The phenomena connected with alternating currents are peculiar and surprising, and I believe there is no person now alive who has anything like the intimate familiarity with their idiosyncrasies than Mr. Oliver Heaviside attained to."

It was, indeed, Heaviside who created basic terminology that became accepted worldwide. In 1884 he introduced the term "impedance," which he defined in a footnote: "Impedance is here and later substituted for apparent resistance. It is the ratio of the amplitude of the impressed force to that of the current when their variations are simple harmonic." The following year he defined resistivity, conductivity, inductivity, and permittivity, and he later coined "capacitance," "inductance," "leakance," and "reactance." In an 1894 review of the first volume of Heaviside's *Electromagnetic Theory*, James Swinburne criticized Heaviside for being "a prolific inventor of new terms" and accused him of "murderous hatred of the Queen's English...."

The first attempt to obtain traces of the waveform of an alternating current was by Jacques Arsène d'Arsonval (1851–1940). D'Arsonval linked a moving coil of the sort used in a galvanometer with a pen that inscribed on a rotating drum. He was principally interested in comparing the physiological effects of alternating current with the effects of faradic current (interrupted direct current). He found a dependence on frequency and on the shape of waveform. By 1888 d'Arsonval, using a Gramme-type generator, had produced frequencies as high as 10,000 cycles per second.

A most effective tool for the study of waveforms came in 1897 with the invention of the cathode ray oscilloscope by Ferdinand Braun (1850–1918), professor of physics at the University of Strassburg. However, almost three decades passed before the oscilloscope became a practical instrument. Braun used a potential of 10,000 to 50,000 volts to accelerate the electrons, which made the tube dangerous, bulky, and expensive. In the early 1920s John B. Johnson of Western Electric designed a tube requiring only 300 to 400 volts. Another difficulty was solved by Frederick Bedell of Cornell University, whose "linear sweep circuit" moved the electron beam across the tube at a constant speed and held a repetitive pattern in place.

We will see in the next section how alternating current soon came to be widely used, principally because electric power could be more economically distributed as alternating current rather than as direct current.

II.8   THE BEGINNINGS OF AC POWER TRANSMISSION

By the 1880s there were effective generators of electricity, especially the dynamos of the Gramme type. Though effective, these machines were not efficient by modern standards. This is not surprising in light of the fact that they were designed without much knowledge of the electromagnetic properties of the materials, nor was the utility of the concept of the magnetic path appreciated. Nevertheless, the principal hindrance to the greater use of electricity was the difficulty of transmitting rather than generating power.

At the 1881 International Electrical Exhibition in Paris several leaders in electrical technology, among them Jacques Arsène d'Arsonval and Marcel Deprez (1843–1918), argued that long-distance transmission of electric power would be economical only at high voltage because of the large amount of copper required to carry the heavy currents at low voltage. Two problems had yet to be solved. The first was how to transform from high voltage for transmission to low voltage for use, and the second was, if alternating current was to be used, how to synchronize two AC generators to run them in parallel.

A solution to the second problem was presented in 1883 by John Hopkinson (1849–1898). In his paper he mentioned in a footnote that Henry Wilde had achieved successful parallel operation of AC alternators as early as 1869. It is interesting that Hopkinson also made important contributions to the scientific design of generators. In 1885 he developed the theory of the magnetic circuit, based on Maxwell's electromagnetic theory. He also drew attention to the leakage of the magnetic field and to the varying magnetic properties of different varieties of iron and steel.

The solution to the first problem—converting from high voltage to low voltage—was, of course, the transformer. It might appear surprising that it took 50 years to reinvent what Faraday had constructed in 1831. It must be remembered, however, that Faraday's iron ring with two coil windings was connected to a battery and that alternating current did not then exist. Faraday was studying induction phenomena, not any transformation of current. Still, the induction coil—which continued to be developed for several purposes, including, as we have seen, medical treatment—was, in a sense, the prototype of the transformer.

Among the first people to understand how to change the voltage of alternating current and to recognize the importance of this in distributing electric power were Marcel Deprez and Jules Carpentier. In 1881 they obtained a patent on a transformer (see Figure II.24). Perhaps more influential, however, was another two-person team: the Frenchman Lucien Gaulard and the Englishman John D. Gibbs. Gaulard and Gibbs applied for patents for transformers to use in their AC system in 1881, and they

Figure II.24 A photograph of a pair of transformers used in AC trials at the Franklin Institute in Philadelphia in 1879 (courtesy of News Bureau, General Electric Company).

demonstrated their use at an exhibition in London in 1883 and at another in Turin in 1884. At Turin they transmitted electric power a distance of 25 miles at 2000 volts, considered high at the time.

In the United States George Westinghouse (1846–1914), who had invented the railway air brake, organized the Union Switch and Signal Company in 1880 and became interested in electric power transmission. Through his advisor, Franklin Pope, he learned of the Turin demonstration by Gaulard and Gibbs. In 1885 Westinghouse acquired the American rights to their patents and assigned an enterprising young electrician, William Stanley (1850–1916), the task of finding how to put the inventions to practical use. In November 1885 Gibbs and Gaulard sent Reginald Belfield with sample transformers to Stanley's laboratory in Great Barrington, Massachusetts. The transformers had iron wire cores and were supposed to be used in series connection to the high-voltage terminals. Stanley and Belfield immediately made improvements and connected the transformers in parallel to eliminate load problems.

Westinghouse also visited a Hungarian electric company, Ganz & Company, in Budapest. There Max Deri, Charles Zipernowski, and Otto Titus Blathy had obtained patents for transformer designs. After investigation, Stanley found them superior in certain respects and partly adopted their designs.

By late March 1886 Stanley had completed construction of an AC transmission system with transformers that raised the voltage to 500 volts. He transmitted electric power to the town of Great Barrington over a distance of 4000 feet, transformed the voltage down to 100 volts and lit up light bulbs in two dentists' offices, the post office, the telephone exchange, and several stores.

On the basis of this success and additional tests in Pittsburgh, Westinghouse in April 1886 organized the Westinghouse Electrical Manufacturing Company. It produced efficient transformers with laminated cores, as well as complete systems of alternating-current equipment. In November 1886 the Westinghouse Company put into operation in Buffalo, New York, its first of many commercial installations.

At this time it was by no means clear that AC systems were preferable to DC systems. Edison, feeling threatened by the competition, steadfastly maintained the superiority of DC equipment and emphasized the greater safety of a system entirely at 110 volts. Edison's campaign against alternating current was vigorous and sustained, reverberating also in Europe.

A more serious hindrance to the adoption of alternating current than the safety issue was the lack of a practical AC motor for industrial application. Since a dynamo can often be operated in reverse as a motor, one might use an alternator as a motor. Such a motor would require, however, that the current supplied and the motor's motion remain exactly in phase. This and other problems—the lack of a starting torque and the variety of frequencies used in AC systems—made the so-called synchronous motor impractical.

A successful AC motor of a different sort, the so-called induction motor, eventually met the need. The idea for this type of motor goes back to Arago's experiment of suspending a magnetic needle above a rotating copper disk. Eddy currents in the disk, induced by the magnet, produce a magnetic field that acts on the magnet. Also, one could induce rotation of the disk by rotating the magnet. It was on this principle that the induction motor was based.

A number of people, notably Walter Bailey in 1879 and Marcel Deprez in 1883, showed that one could obtain a rotating magnetic field by changing the polarity of an array of stationary electromagnets, and in 1885 Galileo Ferraris of Turin constructed an induction motor consisting of a copper cylinder driven by a rotating magnetic field. Others, including O. B. Shallenberger of the Westinghouse Company, built prototypes, but a practical motor design eluded discovery.

The engineer principally responsible for an effective AC motor was Nikola Tesla (1856–1943). A Serbian born in Croatia, Tesla (see Figure II.25) studied in Graz, Austria, and then in Prague. He took a particular interest in alternating-current phenomena, and before his immigration to the United States in 1884 had conceived of a way to produce a rotating magnetic field without mechanical motion and had built an induction motor.

After working for a year for Edison, Tesla established his own laboratory and began filing for patents on his inventions. When Westinghouse learned of his AC motor patents, he bought some patent rights and hired Tesla to work for his company. The

Figure II.25 A photograph of Nikola Tesla (courtesy of the Smithsonian Institution).

hoped-for success with induction motors was not immediately realized, and Tesla left the company in 1889. At about the same time rapid progress was occurring in Europe: Charles E. L. Brown at the Oerlikon Company in Switzerland and Michael von Dolivo-Dobrowolski at the Allgemeine Electrizitäts Gesellschaft (AEG) in Germany also developed induction motors, which became commercially available in 1891.

The first AC power systems transmitted single-phase AC. Some early AC motors, including the Ferraris motor and a Tesla motor, ran on two-phase AC but achieved this by "phase splitting" the single-phase AC supplied by the transmission lines. In the early 1890s some power networks supplied two-phase AC, but what soon became standard—because of steadier power and more efficient transmission—was three-phase AC. By 1890 Tesla had already worked out how to build a complete three-phase system, and Tesla's practical three-phase motor, marketed by Westinghouse, was soon in wide use.

The first major installation of a three-phase electric power system was in 1891 for the Frankfurt Electrotechnical Exhibition. The generating station, in Lauffen on the river Neckar, was connected to the Frankfurt Exhibition by a high-voltage transmission line. This line, which was 110 miles long, transmitted 240 kilowatts at 15 kilovolts. At Frankfurt, transformers reduced the voltage to 220 for use. The generators and transformers were supplied by the Oerlikon Company of Switzerland, and the AC motors were supplied by AEG. This installation was something of a prototype for the gigantic enterprise at the Niagara Falls, which took 13 years—1883 to 1896—from conception to operation; this project involved American as well as European experts under Westinghouse direction and with Oerlikon participation.

Rather than continuing here the account of the growth of electric power networks, we will turn to parallel developments in physics and engineering theory that led to a remarkable blossoming of electrical technology in the twentieth century.

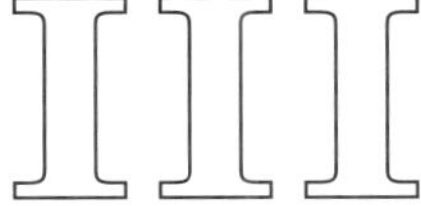

The Great Synthesis

III.1 MATHEMATICAL FOUNDATIONS

An increasing ability to comprehend the world and our place in it is a large part of the epic story of human biological and cultural evolution. The history of science, which documents progress in comprehension over the last two or three millenia, shows that mathematics has assumed a larger and larger role in human knowledge.

The first of the mathematical sciences was astronomy. The Greeks, like the Babylonians before them, made quantitative descriptions of the heavens, specifying positions and times numerically. These efforts drew upon and added to the human repertoire of mathematical techniques. For example, trigonometry was created by Alexandrian Greeks in the second century B.C. for astronomical purposes. Thus, when Hipparchus of Nicaea (ca. 165–127 B.C.) tabulated the length of chords of circles for different angles (see Figure III.1), he was producing the first tabulation of a form of what came to be called the sine function.

The most celebrated of the Alexandrian mathematicians was Claudius Ptolemy (ca. 90–168 A.D.). His 13-book treatment of spherical trigonometry and astronomy is known as the *Almagest* (from the Arabic words meaning the greatest). From Ptolemy's time through the sixteenth century, this was the standard work on astronomy.

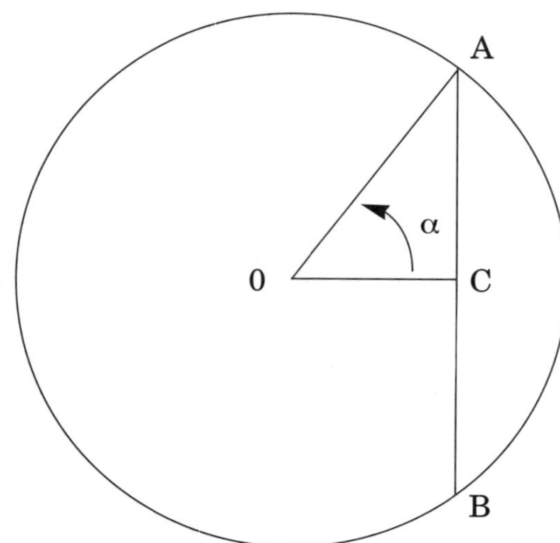

Figure III.1 Hipparchus calculated the length of the chord *AB*
for different values of the angle α with *OA* a fixed
length. This is equivalent to calculating the sine
of α, later defined to be the ratio of *AC* to *OA*.

In 1543 Nicolaus Copernicus (1473–1543) published *On the
Revolutions of the Heavenly Bodies*, which argued that the earth
and the other planets revolve about the sun, but it took a hundred years before most astronomers accepted the heliocentric
model of the solar system. Johannes Kepler (1571–1630) regarded the sun as the center of the solar system, but argued that the
planets follow elliptical orbits. His work won adherents to the
heliocentric model, mainly because he showed that planetary
motions obey three quantitative laws: planets move in elliptical
paths, with the sun at one focus; in the plane of the orbit, a plan-
et sweeps out equal areas in equal times; and the ratio of the
square of the period of revolution to the cube of the mean radius
of the orbit is the same for all planets.

Taking Kepler's laws as a starting point, Isaac Newton
(1642–1727) arrived at a system of mechanics (based on three
laws) and a theory of gravitation (based on a single law of gravitational force), which he presented in the monumental *Mathematical Principles of Natural Philosophy* (1687; see Figure
III.2). At the heart of Newton's achievement was his successful
quantification of the concept of force. Kepler's laws followed as

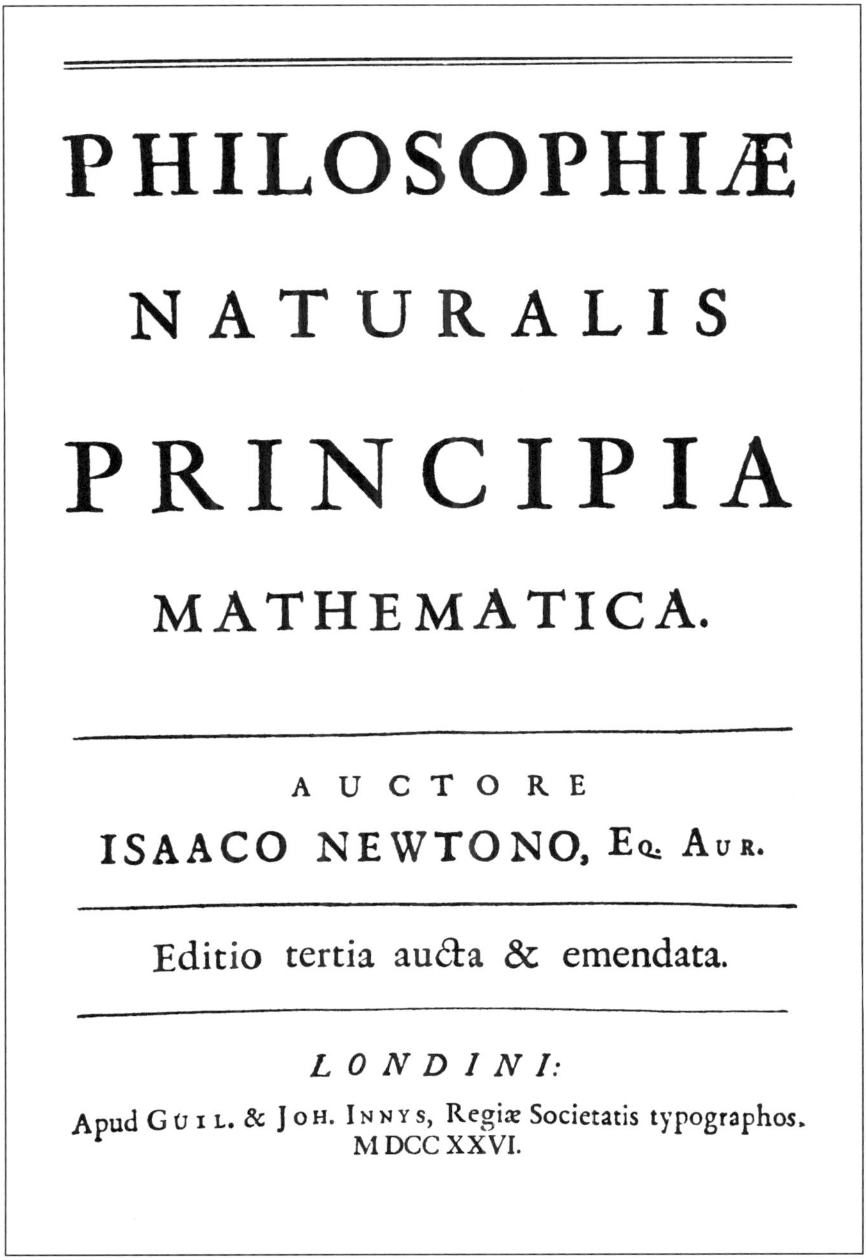

PHILOSOPHIÆ NATURALIS PRINCIPIA MATHEMATICA.

AUCTORE

ISAACO NEWTONO, Eq. Aur.

Editio tertia aucta & emendata.

LONDINI:

Apud Guil. & Joh. Innys, Regiæ Societatis typographos.
MDCCXXVI.

Figure III.2 The title page of the third edition (1726) of Newton's *Principia*.

consequences of Newton's mechanics and the law of gravitation. In the course of this work, Newton developed a new mathematics, what is today called differential and integral calculus.

Calculus became the language for most of the mathematical science of the eighteenth, nineteenth, and twentieth centuries. It was invented independently in Germany. Gottfried Wilhelm Leibniz (1646–1716), who was born in Leipzig and studied in Paris with Christian Huygens, developed his own version of calculus by 1675. It is Leibniz's notation, rather than Newton's, which is today in general use. Several mathematicians on the Continent, notably Jacob Bernoulli (1654–1705) and his brother Johannes Bernoulli (1667–1748), quickly adopted Leibniz's techniques and developed them further. In the eighteenth century the greatest of these was Leonard Euler (1707–1783). Euler made many contributions to calculus, especially to the study of differential equations. His *Introduction to Infinitesimal Analysis*, published in two volumes in 1748, was immensely influential.

Much of the development of differential and integral calculus resulted from efforts to solve problems of physics. One who might equally well be regarded as physicist or mathematician was Joseph Louis Lagrange (1736–1813). His landmark volume *Analytical Mechanics* of 1788 presented for the first time an elegant formulation of the problems of mechanics in terms of kinetic and potential energies. Another Frenchman, Pierre Simon de Laplace (1749–1827), contributed the concept of potential (of a force field) with the classical differential equation in Cartesian coordinates

$$\frac{\partial^2 U}{\partial x^2} + \frac{\partial^2 U}{\partial y^2} + \frac{\partial^2 U}{\partial z^2} = 0.$$

Mathematical analysis reached new heights in Laplace's five-volume treatment of planetary motion, *Celestial Mechanics* (1799–1825).

In the nineteenth century mathematics continued to be advanced by people seeking to understand the physical world. Extraordinarily useful mathematical techniques were developed by Joseph Fourier (1768–1830) in his study of the flow of heat in solid bodies. Fourier's *Analytic Theory of Heat* (1822) introduced the concept of boundary value problems and demonstrated, without proof, the representation of given functions by means of trigonometric series. The latter is today called Fourier series representation and is used in most branches of science and engineering. Extending the work of Fourier, George Green (1793–1841) and George Gabriel Stokes (1819–1903) developed theorems for the solution of three-dimensional boundary value problems. All of them received their stimulus for mathematical invention from attempts to solve problems of physics. Fourier wrote

> The profound study of nature is the most fecund source of mathematical discoveries. Not only does this study, by offering a definite goal to research, have the advantage of excluding vague questions and futile calculations, but it is also a sure means of molding analysis itself and discovering those elements in which it is essential to know and which science ought always to conserve. These fundamental elements are those which recur in all natural phenomena.

Perhaps the greatest mathematician of the nineteenth century, Carl Friedrich Gauss (1777–1855), shared this view. At the age of 22 he proved the fundamental theorem of algebra: a polynomial equation with complex coefficients of order n has n complex roots. (Gauss showed, moreover, that for a polynomial with real coefficients, the nonreal roots occur as conjugate pairs, hence the polynomial can be factored into real linear and quadratic factors.) Besides his purely mathematical work, Gauss made important contributions to astronomy, electrostatics, geodesy, and the study of terrestrial magnetism. The so-called theorem of Gauss (or divergence theorem) is fundamental to gravitational theory and to electrostatics (where it asserts that the integral of the electric intensity over any closed surface equals the net charge within the surface).

Many other mathematicians took part in the great development of mathematical analysis in the nineteenth century. Augustin-Louis Cauchy (1789–1857) defined the concept of an analytic function of a complex variable, and he established the Cauchy integral relation (which states that any line integral of a closed, analytic region is zero). Karl Theodor Wilhelm Weierstrass (1815–1897) extended the theory of functions of a complex variable and studied periodic functions of higher order and the convergence of infinite series. Georg Friedrich Bernhard Riemann (1826–1866) made an important contribution in his dissertation, completed at the age of 25, on "Foundations for a general theory of functions of a complex variable." Riemann also introduced conformal mapping and what is now called Riemann integration in the complex plane for many-valued functions. He was keenly interested in physical problems and made major contributions to the theory of dynamics. And Henri Poincaré (1854–1912), who has been called "the last universalist," enriched the theory of functions by introducing the class of automorphic functions.

In the first decades of the twentieth century the trend toward a separation of mathematics and physics became noticeable, and the leading mathematicians were seldom concerned with the applications of mathematics. In 1924 Richard Courant (1888–1972) wrote

> Since the seventeenth century, physical intuition has served as a vital source for mathematical problems and methods. Recent trends and fashions have, however, weakened the connection between mathematics and physics; mathematicians, turning away from the roots of mathematics intuition, have concentrated on refinement and emphasized the postulational side of mathematics ...

Fortunately for the science of electrical engineering, the mathematics that had already been developed was of great utility in practical applications.

III.2 THE INTERNATIONAL SYMBIOSIS

Electrical engineering has been a collective enterprise, remarkable for its continued development over more than a century and a half, remarkable for the depth and range of its study and application, and remarkable for the degree of international collaboration. It is the purpose of this section to illustrate these remarkable features in three charts. The first of these, Table III.1, lists some of the great contributors to electrical engineering, arranging them chronologically in each of four categories.

The first category is mathematics, and the people listed there made contributions to mathematics that turned out to be important in electrical engineering. This list begins with Newton and Leibniz, and their names are emphasized because the differential and integral calculus, which they invented, is fundamental to almost all mathematical physics. Here, as in the other lists, we have selected only a few of the most illustrious contributors.

The second category is electrophysics, or the physics of electromagnetic phenomena. The people listed here were seeking new knowledge. William Gilbert long preceded the others. He and many of the others up to and including Faraday were not seeking knowledge in a mathematical form; after Faraday the contributions were highly mathematical. Faraday's and Maxwell's names properly bear emphasis, the former for revealing many of the basic phenomena of electromagnetism and for introducing the concept of a field connecting actions and properties through space, the latter for devising an encompassing mathematical theory of electromagnetism. This is, to be sure, a very incomplete listing, but it does suggest the wealth of contributions over more than two centuries, from electrostatic machines to quantum electronics.

The fourth category is invention. Here are listed a few of the creative people who have found practical uses for electromagnetic phenomena. Edison's name is emphasized because of the range and success of his inventive work. Many of the early inventors, including Edison, had little scientific background, and often the devices introduced by such people underwent considerable improvement when scientific understanding was applied to the design of the devices. Some of the inventors listed here, such as John Ambrose Fleming and Vladimir Zworykin, had considerable scientific training, which they drew upon in their work.

With the development of mathematical theory, there was more knowledge that might be applied to practical concerns, but at the same time the mathematical sophistication made theory less accessible to engineers. At the turn of the century, most electrical engineers, even most teachers of electrical engineering, did not understand the work of Maxwell or Helmholtz. Crucial to the exploitation of theory was the work of "translators," mathematically gifted electrical engineers such as Oliver Heaviside and Charles Steinmetz. This is the third category. In this listing, Heaviside's name properly bears emphasis, as his three-volume *Electromagnetic Theory* became the prototype of the application of physical theory to the problems of electrical engineering.

Table III.1 does give something of a chronological overview, but the actual spacing in time is better indicated in Table III.2, which shows the lifespans of the contributors on a time line on which are marked a dozen milestones in the history of electrical technology. Looking at Table III.2, one can imagine a flow from mathematical development, through physics and EE science, to implemented technology. Though this flow does not account for most of electrical technology, it has undoubtedly occurred and can be traced for many of the technological advances of the twentieth century.

By classifying the names listed in the two first tables according to nationality, one may obtain Table III.3 (where the names are placed chronologically in each column). Its purpose here is to emphasize the international nature of the development of electrical engineering. Of course, if names were added other nationalities would be represented and still no nationality would dominate. In this respect, electrical engineering resembles almost all fields of science, where the free interchange of results of research—both theoretical and experimental—has been the rule rather than the exception. The United States has benefited from its encouragement of immigration, as is shown by the large number of U.S. contributors in Table III.3 who were first-generation U.S. citizens.

In the next section we will review the work of James Clerk Maxwell, whose brilliant theory of electromagnetism consolidated and advanced preexisting knowledge and predicted the existence of electromagnetic waves.

III.3 MAXWELL'S ELECTROMAGNETIC THEORY

The discoveries of Volta, Ampère, Ohm, Faraday, and others led to the new technologies of telegraphy, electrochemistry, and arc lighting. These discoveries also helped build what was by midcentury a rather sophisticated electrical science. There was,

Table III.1 Leading contributors to electrical engineering in the areas of mathematics, electrophysics, EE science, and invention

Mathematics	Electrophysics	EE Science	Invention
	William Gilbert 1544–1603		
Isaac Newton 1642–1727	Benjamin Franklin 1706–1790		
G. W. Leibniz 1646–1716	Luigi Galvani 1737–1798		
Leonhard Euler 1707–1783	Alessandro Volta 1745–1827		
J. B. J. Fourier 1768–1830	A. M. Ampère 1775–1836		
C. F. Gauss 1777–1855	H. C. Oersted 1777–1851		
A. L. Cauchy 1789–1857	G. S. Ohm 1787–1854		
George Green 1793–1841	**Michael Faraday** 1791–1867		Samuel Morse 1791–1872
N. H. Abel 1802–1829	Joseph Henry 1797–1878		Charles Wheatstone 1802–1879
George Boole 1815–1864	Wilhelm Weber 1804–1891		Werner von Siemens 1816–1892
K. Weierstrass 1815–1897	Hermann Helmholtz 1821–1894		Z. T. Gramme 1826–1901
G. F. B. Riemann 1826–1866	William Thomson 1824–1907		Antonio Pacinotti 1814–1912
Henri Poincaré 1854–1912	**James Maxwell** 1831–1879		G. Westinghouse 1846–1914
David Hilbert 1862–1943	Wilhelm Röntgen 1843–1923		A. G. Bell 1847–1922
Henri Lebesgue 1875–1941	Oliver Lodge 1851–1940		**Thomas Edison** 1847–1931
	J. H. Poynting 1852–1914		John A. Fleming 1849–1945
	J. J. Thomson 1856–1940	**O. Heaviside** 1856–1925	Nikola Tesla 1856–1943
	Heinrich Hertz 1857–1894	Michael Pupin 1858–1935	Paul Nipkow 1868–1940
	Max Planck 1858–1947	A. E. Kennelly 1861–1939	Lee de Forest 1873–1961
	Albert Einstein 1879–1955	Charles Steinmetz 1865–1923	Guglielmo Marconi 1874–1937
	Niels Bohr 1885–1964	G. A. Campbell 1870–1954	Vladimir Zworykin 1889–1982
	E. Schrödinger 1887–1961	Irving Langmuir 1881–1957	E. H. Armstrong 1890–1954
	William Bragg 1890–1971	K. W. Wagner 1883–1913	Philo Farnsworth 1906–1971
	W. Heisenberg 1901–1976	Dennis Gabor 1900–1979	

Table III.2   A time line showing the lifespans of leading contributors to electrical engineering as well as particular events in the history of electrical technology

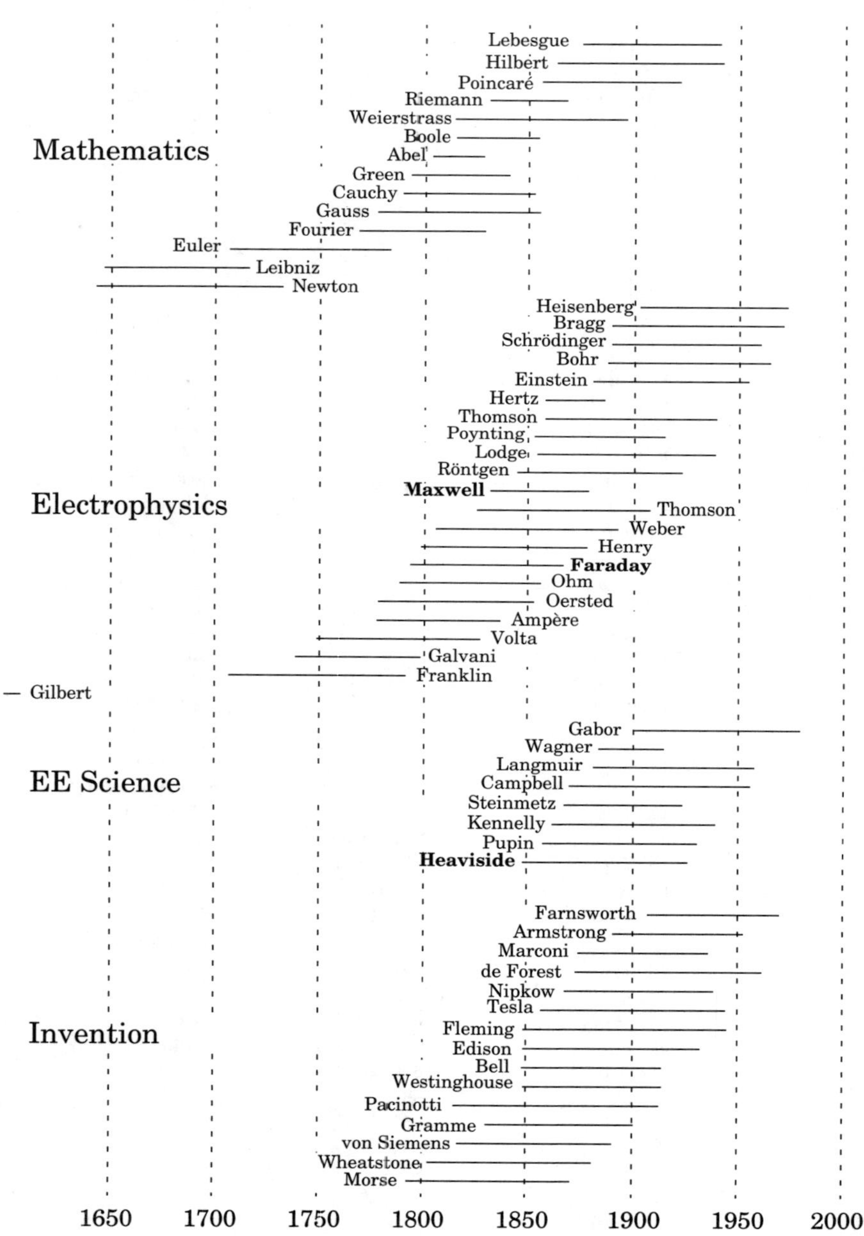

Table III.3   Leading contributors to electrical engineering classified according to nationality

British	French	German	U.S.	Other
Gilbert	Fourier	Leibniz	Franklin	Galvani–Italian
Newton	Ampère	Euler	Morse	Volta–Italian
Faraday	Cauchy	Gauss	Henry	Oersted–Danish
Green	Gramme	Ohm	Westinghouse	Abel–Norwegian
Wheatstone	Poincaré	W. Weber	Bell*	Pacinotti–Italian
Boole	Lebesgue	Weierstrass	Edison	Marconi–Italian
W. Thomson		von Siemens	Tesla*	Bohr–Danish
Maxwell		Helmholtz	Pupin*	
Fleming		Riemann	Kennelly*	
Lodge		Röntgen	Steinmetz*	
Poynting		Hertz	Campbell	
Heaviside		Planck	de Forest	
J.J. Thomson		Hilbert	Langmuir	
Bragg		Nipkow	Zworykin*	
		Einstein	Armstrong	
		Wagner	Gabor*	
		Schrödinger	Farnsworth	
		Heisenberg		

*First-generation U.S. citizen.

however, no general theory that encompassed all the known electrical and magnetic phenomena. Most scientists sought an action-at-a-distance theory, similar to Newton's extremely successful theory of gravity. The nonmathematical Faraday had a different conception. As we saw in Section II.3, he imagined electric and magnetic lines of force that extended through space, and the magnetic lines he made visible with iron filings. The person who turned Faraday's conception into an elegant and comprehensive mathematical theory was James Clerk Maxwell (1831–1879). (See Figure III.3.)

Maxwell was born in Edinburgh and studied at the University of Edinburgh and then Trinity College, Cambridge, from which he graduated in 1854. His interest in Faraday's experiments was great, and in 1855 he wrote a paper on Faraday's lines of force, describing them mathematically as tubes starting at positive charge areas and terminating at areas of equal amounts of negative charges. He also expressed mathematically the magnetic field lines surrounding currents. The highly mathematical presentation moved Faraday to say in an 1857 letter to Maxwell: "When a mathematician engaged in investigating

Figure III.3 A portrait of James Clerk Maxwell.

physical actions and results has arrived at his conclusions, may they not be expressed in common language as fully, clearly, and definitely as in mathematical formulae?"

In 1856 Maxwell was appointed to the chair of natural philosophy at Marischal College, Aberdeen. He made a theoretical study of the rings of Saturn, showing that they must be composed of loose particles since solid rings would not be mechanically stable. He elaborated the kinetic theory of gases, publishing in 1860 a law of the distribution of molecular energies, now known as the Maxwell-Boltzmann law. And he investigated the superposition of colors, showing that any color can be composed of three basic colors, scarlet, green, and blue. (He gave their wavelengths in millionths of the Paris inch, the unit customary at the time.) For this work the Royal Society awarded him the Rumford Medal.

From 1860 to 1865 he served as professor of physics and astronomy at King's College, London. There he turned again to electromagnetics, and in 1861 he elaborated the notion of a force field. He introduced the concept of the "displacement current," the time variation of the electric field, which has the character of time rate of change of charge and thus the same as an electric current. This idea may have been suggested by Faraday's notion of stress in the dielectric when in the "electrotonic" state.

It was in this 1861 paper that Maxwell asserted that light is an electromagnetic phenomenon. He had earlier been impressed by Faraday's discovery of the rotation of polarization of light by a magnetic field. In this paper he calculated the speed of transverse waves in the hypothesized electromagnetic medium and found that this velocity "agrees so exactly with the velocity of light...that *we can scarcely avoid the inference that light consists in the transverse undulations of the same medium which is the cause of electric and magnetic phenomena*" [emphasis in original].

It was about this time that Maxwell became involved with an effort to devise practical units for the measurement of electric and magnetic quantities. In 1861 the British Association for the Advancement of Science (BAAS) appointed a committee of scientists to study the issue. The committee, which included William Thomson, Charles Wheatstone, and H. C. F. Jenkin (1833–1885), sought in particular to establish a standard of resistance related to the absolute system of units, the cgs (centimeter, gram, second) system initiated by Gauss and expanded by Wilhelm Weber.

A provisional report of that committee was published in 1862, which accepted the procedure proposed by Weber for the absolute measurement of magnetic field intensity and the procedure proposed by Maxwell for defining electrostatic units and electromagnetic units. Maxwell suggested that Coulomb's law

$$F = \frac{q_1 q_2}{d^2},$$

with cgs units for force and distance, be used to define a basic
unit of charge, and that a force law for (fictitious) magnetic poles

$$F = \frac{m_1 m_2}{d^2},$$

again with cgs units, be used to define a basic magnetic unit (the
gauss).

In 1863 the BAAS published a further report on progress on
electrical and magnetic standards. Maxwell and Jenkin proposed
for the practical unit of resistance the name "ohm," defining its
value as 10^8 absolute units. As a result of their participation in
the searching discussions on units and measurements, Maxwell
and Jenkin published an exposition of this work in 1865. Maxwell's work on standards was also reflected in his continuing concern with dimensional analysis (a method of analysis that
expresses all quantities involved in terms of the fundamental
units), a method he was one of the first to use to great advantage.

We may point out, as an aside, that the BAAS was also seeking international agreement on practical units of measure. In
1862 it invited suggestions from such leaders of electrical science
as Joseph Henry in Washington, D.C., E. Edlund in Uppsala, C.
Matteucci in Turin, Moritz von Jacobi in St. Petersburg, C. S. M.
Pouillst in Paris, Franz Neumann in Königsberg, J. C. Poggen-
dorff and Werner von Siemens in Berlin, and Wilhelm Weber in
Göttingen. Apparently stimulated by this step toward international interchange, the French government arranged for a "Conference Diplomatique du Mètre" in Paris at which 20 nations
participated. There it was agreed to establish an international
organization for the comparison of standards as well as for the
custody of prototypes. Thus came into existence the International Bureau of Weights and Measures at Sèvres near Paris in
1875. In 1881 the first International Congress of Electricians,
held in Paris at the time of the International Electrical Exhibition, adopted the recommendations of the BAAS committee on
standards, namely, the "absolute practical system" of units
based upon the cgs system but extended to include the ohm as
the unit of resistance. The unit of current, the ampere, was defined as the current produced by one volt applied to a resistance
of one ohm; the unit of charge, the coulomb, was defined as the

charge accumulated by one ampere flowing for one second; and the unit of capacitance, the farad, was defined as that of a condenser holding one coulomb at one volt. The need to verify standards led to the establishment of the Physikalisch-Technische Reichsanstalt (Imperial Physical-Technical Laboratory) in Germany in 1887, the National Physical Laboratory in England in 1889, and the National Bureau of Standards in the United States in 1901.

In late 1864 Maxwell completed a paper that for the first time presented his remarkable synthesis of what had been learned about electromagnetism. Entitled "A dynamical theory of the electromagnetic field," it was read at a meeting of the Royal Society on December 8 and was published in the Society's *Philosophical Transactions* the following year. In this paper he stressed the field aspects of electromagnetism and argued that the "vortices" of a changing magnetic field would link with "vortices" of the displacement current, which would in turn link with other "vortices" of the magnetic field. In this way an electromagnetic field can be propagated through space, and Maxwell derived the wave equation for a plane electromagnetic wave with sinusoidal time variation.

In free space, where charge density and conductivity are everywhere zero, Maxwell's equations assume a simpler form. For an electric field in the x-direction and a magnetic field in the y-direction, we have the pair of equations

$$\frac{\partial H_y}{\partial z} = \varepsilon \frac{\partial E_x}{\partial t}, \qquad \frac{\partial E_x}{\partial z} = -\mu \frac{\partial H_y}{\partial t}.$$

By differentiating the first with respect to t and the second with respect to z, and then adding the equations, one obtains what some may recognize as a second-order wave equation:

$$\frac{\partial^2 E_x}{\partial z^2} - \mu\varepsilon \frac{\partial^2 E_x}{\partial t^2} = 0.$$

If c is defined to be the quantity

$$\frac{1}{\sqrt{\mu\varepsilon}},$$

a solution to the original pair of equations is

$$E_x = A \cos\omega\left(t \pm \frac{z}{c}\right), \qquad H_y = \sqrt{\frac{\varepsilon}{\mu}}A \cos\omega\left(t \pm \frac{z}{c}\right),$$

where A is an arbitrary amplitude and ω is an arbitrary angular frequency. This solution is a sinusoidal wave traveling at speed c, where c is easily calculated, since m and e are measurable quantities. The fact that this quantity turned out to equal, almost exactly, the experimentally determined speed of light, Maxwell regarded as strong evidence that light itself was an electromagnetic wave.

In 1865 Maxwell resigned his professorship at King's College and retired to the family's estate Glenlair in Kirkenbright, Scotland. There he continued his reflections on electromagnetism, working to compose a systematic presentation of the subject. In 1871 Cambridge University established a new chair in experimental physics and created the Cavendish Laboratory. Friends of Maxwell persuaded him to accept the chair and the position of director of the laboratory. [Later Maxwell collected and edited the writings of Henry Cavendish (1731–1810), a pioneering experimenter, most of whose results were then still unpublished.]

Maxwell was concerned also with thermodynamics, especially the kinetic theory of gases. His work in this area, collected in two treatises, *Theory of Heat* published in 1871 and *Matter and Motion* published in 1876, helped establish a new approach to thermodynamics, what came to be called statistical mechanics. In 1873 his great work on electromagnetism, *Treatise on Electricity and Magnetism*, appeared in two volumes (see Figure III.4). The significance of this work was not widely appreciated at the time, partly because of its mathematical inaccessibility, partly because a Newtonian approach to the subject was still the norm, and partly because experimental results did not at that time show Maxwell's theory to be superior to rival theories.

Maxwell died in 1879, at the age of 48, before the work of Heinrich Hertz confirmed his prediction of electromagnetic waves and before the further development of his theory in the hands of Oliver Heaviside and others. Today, of course, Maxwell's theory is regarded as one of the greatest achievements of

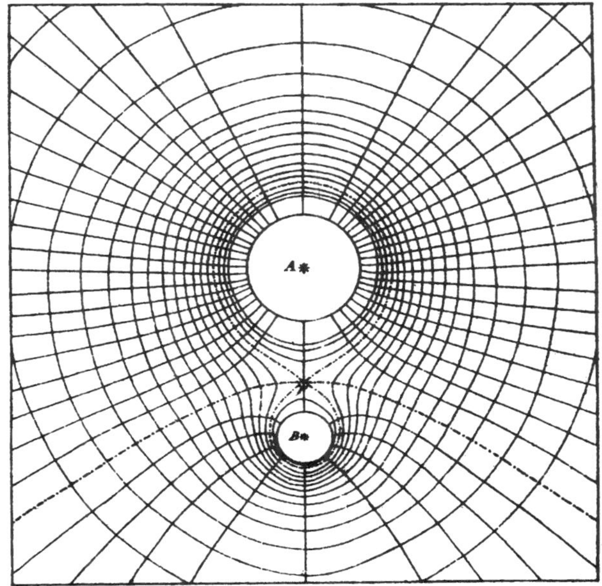

Figure III.4 An illustration from Maxwell's "Elementary treatise on electricity" showing lines of force and equipotential surfaces.

the human intellect. Einstein called it "the first great fundamental advance in theoretical physics since Newton." The physicist Richard Feynman wrote:

> From a long view of the history of mankind—seen from, say, ten thousand years from now—there can be little doubt that the most significant event of the nineteenth century will be judged as Maxwell's discovery of the laws of electrodynamics. The American Civil War will pale into provincial insignificance in comparison with his important scientific event of the same decade.

The Beginning of Scientific Electrical Engineering

IV.1 Oliver Heaviside and the Application of Maxwell's Theory

Perhaps no personality among engineers has provoked more discussion and controversy than Oliver Heaviside, yet no one has made a more significant contribution to the teaching of Maxwell's electromagnetism than he did (see Figure IV.1). Heaviside was born May 13, 1850, and as a nephew of Sir Charles Wheatstone, it was probably natural that he started as a telegraph operator. In 1868 he began work for a private Danish-English telegraphy company in Denmark. When the British Post Office took over telegraphy in 1870, Heaviside was transferred to the office at Newcastle-on-Tyne, where he became chief operator in 1871.

While working as a telegrapher, Heaviside pursued his own study of the science of electricity. Demonstrating considerable mathematical ability, he began to publish papers on telegraphic circuits, notably duplex circuits, which allowed messages to be sent in both directions simultaneously. Heaviside saw Maxwell's *Treatise on Electricity and Magnetism* in a library shortly after it appeared in 1873, and he decided at once to study this work. It fascinated him because of its far-ranging new conceptions of electromagnetism, and in May 1874 he gave up his employment and concentrated on studying the treatise.

Heaviside learned also from the writings of William Thomson (later Lord Kelvin). Heaviside verified Thomson's theory of

Figure IV.1 A photograph of Oliver Heaviside.

the submarine cable that led to the K-R law mentioned in Chapter II, and, adding consideration of inductance of the cable, modified the differential equation to arrive at a pair of equations that indicated the possibility of oscillatory behavior, which in fact had been observed. He also read Thomson's paper on the double-energy circuit, and it directed his attention to the general problem of electric propagation along transmission lines. One of his next papers examines the idealized case of lossless lines and demonstrates the infinite sequence of reflections propagating on the finite line, the solution being given in terms of traveling waves.

Apparently on the basis of these publications, the editor of the technical journal *The Electrician*, Charles Henry Walker Biggs, in 1882 asked Heaviside for contributions to his journal. Thus began a series of articles on Maxwell's electromagnetic theory, starting with "The relations between magnetic force and electric current." These articles employed vector notation and presented Maxwell's field equations in their simplest, most elegant form:

$$\text{curl } \mathbf{B} = \mu\left(G + \frac{\partial D}{\partial t} \right) \qquad \text{div } \mathbf{B} = 0$$

$$\text{curl } \mathbf{E} = \frac{-\partial B}{\partial t} \qquad \text{div } \mathbf{D} = \rho.$$

$\mathbf{B}$ is the vector of magnetic induction (magnetic flux density), $\mathbf{G}$ is the vector of electric current density, $\mathbf{D}$ is the electric displacement vector, $\mathbf{E}$ is the vector of electric field strength, ρ is the static charge density, and μ is the magnetic permeability. These so-called duplex equations explain the world of electromagnetism in mathematical terms, an achievement that is certainly akin to Newton's law of gravitation, both in elegance and in power.

Heaviside then added sections in *The Electrician* series on the electromagnetic energy flow. In this investigation he admitted the priority of John Henry Poynting (1852–1914), and today the vector representing energy flow is called the Poynting vector. Heaviside also explored the current distribution in wires with alternating currents, and he evaluated the skin effect. In June 1884, in his study of alternating currents, Heaviside introduced the concept of "impedance" for the apparent resistance of an alternating-current circuit. In the same article he discussed the production of eddy currents by alternating magnetic fields.

In January 1885 Heaviside wrote the first of an important group of articles in the continuing series of *The Electrician* articles. It was entitled "Electromagnetic induction and its propagation" and developed the four-parameter theory of electrical transmission lines. This led to his concept of the "distortionless transmission," which is a relation among the four parameters per unit length—line resistance r, inductance l, capacitance c, and leakance g—namely,

$$\frac{r}{l} = \frac{g}{c} \ .$$

Since leakance g generally is intentionally kept very low, this relation indicates that an increase in inductance l would be needed to achieve faithful signal transmission, a suggestion that was initially fought tooth and nail, since inductance was then usually related only to the slow buildup of current. Heaviside, in fact, had experimented—with poor results—with mixing iron powder into the insulation in order to increase inductance. Heaviside, however, became very combative toward critics of his work, especially Sir William Henry Preece (1834–1913), head of the British Telegraph and Telephone System.

Heaviside felt that mathematically demonstrated truth should take precedence over any so-called "practical" consideration. The editor of *The Electrician*, C. H. W. Biggs, tried to tone down Heaviside's verbal attacks, but with little success, and on October 14, 1887, he resigned because of ill health. The new editor, former assistant to Biggs, fared no better, and on November 30, 1887, he informed Heaviside of the discontinuance of his article series. Fortunately, already in August 1886 Heaviside had started a series of articles in the *Philosophical Magazine*, so he was able to continue to publish his work.

In 1884 Heaviside, in *The Electrician*, initiated the use of what he called "operational calculus." He had observed that differentiation and integration of exponential functions of argument pt, with t signifying time, resulted in multiplication by powers of p. This can be used to "algebraize" calculus. He named p the "operator" and for linear circuit arrangements the impedance function $Z(p)$ became a rational polynomial for which he developed a method for expansion into partial fractions. He followed this in the next section by repeating his exposition of "The distortionless telegraph circuits" from *The Electrician*, restating the relation given that showed the value of substantially increased inductance per unit length of line.

There followed in early 1888 an extensive series of articles on "electromagnetic waves," expounding Maxwell's theoretical structure in Heaviside's manner using the two duplex equations and stressing wire conductors as guides for waves progressing in unbounded space. He also treated reflection in finite lines and the theory of light as electromagnetic waves.

These scientific presentations of Maxwell's theory and its application to special cases with explicit solutions finally impressed William Thomson. At his election to the presidency of the Institution of Electrical Engineers on January 10, 1889, he stated in his presidential address that the theory of wave propagation along wires

> has been worked out in a very complete manner by Mr. Oliver Heaviside; and Mr. Heaviside has pointed out and *accentuated this result of his mathematical theory that electromagnetic induction is a positive benefit*; it helps carry the current. It is the same kind of benefit that mass is to a body shoved along a viscous resistance...Heaviside's way of looking at the submarine cable problem is just one instance of how the highest mathematical power of working and judging as to physical application helps on the doctrine, and directs it into a practical channel.

Heaviside acknowledged this praise promptly, feeling much gratified to have his contributions recognized.

In the fall of 1889, Heaviside moved with his parents from London to Paignton, near the resort town of Torquay, into the house of his older brother Charles. This initiated an easier period of life for him. A change occurred also with *The Electrician*. The editor, Snell, died in March 1890 and on October 23, 1890, the new editor, Alexander Pelham Trotter (1857–1947), invited Heaviside to resume publishing in that journal. Thus, on January 2, 1891, Heaviside started with his exposition of Maxwell's electromagnetic theory in vector notation, and this found very gratifying response internationally. Of course, the strongest endorsement of Maxwell's concepts—and thus of Heaviside's missionary work—had come from the experimental demonstration of electromagnetic waves through the work of Heinrich Hertz (1857–1894) in Karlsruhe in 1888.

Because of the value of Heaviside's contributions to *The Electrician* over the years 1882 to 1887, Macmillan acquired the rights to publish the collection in two volumes, which appeared in 1892. Unfortunately, the readership of scientific literature at that time was very limited, so that the expected remuneration did not result, and this caused concern to Heaviside's friends. It had been agreed also to publish in book form the new series in *The Electrician*. This resulted in three volumes, entitled *Electromagnetic Theory*, the first of which appeared in 1893.

In June 1891 Heaviside was elected a fellow of the Royal Society with the citation "Learned in the science of Electromagnetism, having applied high mathematics with singular power and success to the development of Maxwell's theory, of electromagnetic wave propagation, and having extended our knowledge of facts and principles in several directions and into great detail." The Certificate of a Candidate for Election was signed by, among others, Oliver Lodge, William Thomson, George Francis Fitzgerald, and John Poynting. Heaviside certainly felt this a great tribute.

In early 1893 Heaviside, as fellow of the Royal Society, submitted Part I of a treatise "On operators in physical mathematics" for publication in the Society's *Proceedings*. Traditionally, papers submitted by fellows were published without review, but when Part III was submitted, there was a call for review and the referee rejected publication because of lack of mathematical rigor. Indeed, Heaviside had used "loosely divergent" series expansions for which Thomas John l'Anson Bromwich (1875–1929) more than 20 years later gave a rigorous proof in terms of Riemann integration in the complex plane.

Heaviside recognized that not all his derivations were mathematically rigorous, and upon receiving the letter by Lord Rayleigh, who acted as secretary of the Royal Society, he withdrew Part III. He later included his treatise in abbreviated form as Chapter VIII in the second volume of the book, *Electromagnetic Theory,* under the title, "Generalized differentiation and divergent series."

Though a difficult personality with heightened sensitivities, Heaviside did great service as the missionary of Maxwell's theory of electromagnetism and a pioneer in the practical application of mathematical theory. He died in Torquay on February 3, 1925, practically a recluse. Sir Oliver Lodge, who was a close friend, wrote an extensive memorial sketch in which he said:

> Occasionally a remarkable genius arises, we know not how or whence. Such a one runs the risk of being misunderstood and ne-

glected in his lifetime—partly because he has not gone through the ordinary processes of education, and has therefore not attracted the attention of his contemporaries in the regular way, and partly because his insight may be of the unorthodox and exceptional type and his utterances personal and peculiar in style. He is also apt to be more or less in advance of his time, so that it is sometimes left to posterity to discover the brilliance and notable character of his achievement…such a man was Oliver Heaviside…

IV.2 University Departments of Electrical Engineering

During the Renaissance, universities had faculties of theology, medicine, law, and philosophy. What we would today consider physics was regarded as part of natural philosophy, and the latter did not involve much mathematics before the publication in 1687 of Newton's monumental *Principia* (whose full title is *Philosophiae naturalis principia mathematica* or *Mathematical Principles of Natural Philosophy*).

The distinction between engineering education and physics education goes at least as far back as the founding in France in 1747 of the École des Ponts et Chaussées (School of Bridges and Roads). The scientific basis of material designs was clearly recognized with Napoleon's establishment in 1797 of the École Polytechnique, which came to have Lagrange, Fourier, Ampère, and other leading mathematicians and physicists as teachers.

In the United States, engineering education began with the establishment of the military academy at West Point, New York, in 1802. The first U.S. school of civil engineering (in contrast to military engineering), Rensselaer Polytechnic Institute, was founded in 1824 and awarded its first degree in civil engineering in 1835. Union College followed in 1845, Polytechnic Institute of Brooklyn in 1854, Cooper Union in 1859, and Massachusetts Institute of Technology in 1861. Some universities organized colleges of engineering, which they called "Scientific Schools": Sheffield at Yale and Lawrence at Harvard, both in 1847, and Chandler at Dartmouth in 1851. In 1862 Congress passed the Morrill Act "donating public lands to the several states and territories which may provide colleges for the benefit of Agriculture and the Mechanic Arts," and this led to the creation of the land-grant colleges in each state. Partly as a result of this law, the number of engineering colleges increased to more than 80 by 1880. Among the new institutions was Cornell University, founded by Ezra Cornell in 1865.

The establishment of the École Polytechnique in Paris apparently stimulated the creation in Germany of Technische Hochschulen (now called technical universities). The first was established in 1825 in Karlsruhe, and this was where Heinrich Hertz demonstrated the existence of electromagnetic waves in the late 1880s. Other cities that gained technical universities were Berlin-Charlottenburg, Munich, and Breslau. Similarly, technical universities were created in Zurich—this the "ETH," the Eidgenössische Technische Hochschule, which soon became world famous—and in Vienna.

It was later that an engineering school was established in England: the first polytechnic institute there was founded by Quintin Hogg in 1882. In England such institutions for technical education were generally considered inferior to universities. (This attitude contributed to the gap between the humanities and the sciences deplored by C. P. Snow in his book, *The Two Cultures*.) The first chair of engineering in any British university was created in Glasgow, Scotland, in 1840, and Cambridge University had no chair of engineering until 1875.

Electricity and magnetism are phenomena not directly accessible to our senses, which partly explains why they were discovered rather late in our interaction with the outside world and why their interrelationship needed to be formulated mathematically. Up until the late nineteenth century, the science and application of electricity were almost everywhere taught as part of physics. In 1882, upon the urging of Werner von Siemens, the technical universities in Stuttgart, Darmstadt, and Berlin-Charlottenburg created electrical engineering departments. In 1883 the technical universities in Aachen, Braunschweig, Breslau, Karlsruhe, Dresden, Zurich, and Vienna followed suit.

In the United States the first option of electrical engineering within a physics department occurred at the Massachusetts Institute of Technology under Professor Charles Cross in 1882; two years later this became the Electrical Engineering Program or

Course VI. An electrical engineering department was established at Cornell University in 1883 under Professor William Anthony (1835–1908), who had just returned from Europe. In 1883 the physics department of Lehigh University in Bethlehem, Pennsylvania, became the Department of Physics and Electrical Engineering. In 1886 Polytechnic University of Brooklyn established the Department of Applied Chemistry and Electrical Engineering under Dr. Samuel Sheldon, and the first B.S. in electrical engineering was awarded in 1891. The University of Wisconsin and Stanford University formed Departments of Electrical Engineering in 1891 and 1892, respectively. Subsequently many of the larger universities, including Princeton, Yale, and Harvard, established electrical engineering degree programs.

Besides the fact that many EE programs originated in physics departments, there was the example of Heaviside's fruitful application of mathematics and physics to practical engineering. So it is not surprising that by the turn of the century many EE programs emphasized physics and mathematics. The physical and mathematical basis of electrical engineering was revealed in Cyprien O. Mailloux's 1891 tribute to Professor William Anthony "as an Electrician and as a Physicist": "We are at last entering on a phase where the electrical engineer is *superseding* the inventor, where the necessity for paying strict attention to electrical and mechanical engineering requirements is becoming obvious to us all."

Along with the new courses in electrical engineering came the publication of textbooks. One of the first texts was by Gisbert Kapp (1852–1922): *Electric Transmission of Energy and Its Transformation, Subdivision, and Distribution*, first published in 1883 (subsequent editions appeared in 1886 and 1891). Kapp, who was born in Austria, became the first professor of electrical engineering at the University of Birmingham. In this book he included a history of machinery and of electrical units, both absolute and practical. In 1885 Kapp complained that "nothing has yet been published on the construction of dynamos, which would be of practical value to the manufacturer. Of theories there are more than enough, but the connecting links between pure science and practical work are still missing." (It is well to remember that the great battle of DC versus AC was still raging, though the scales were pointing heavily in favor of AC.)

In 1890 Arthur Edwin Kennelly and H. D. Williamson wrote *Practical Notes for Electrical Students* as a study aid for the new field of alternating currents. This book used phasor graphs to show the phase relationships in Argand diagrams. Silvanus P. Thompson followed in 1892 with a textbook entitled *Dynamo-Electric Machinery*. This included a clear presentation of the magnetic circuit, based upon Hopkinson's earlier publication.

The most important contributor to alternating-current theory was Charles Proteus Steinmetz (1865–1923). He studied electrical engineering at the University of Breslau (now Wroclav), but left Germany for political reasons in 1889. He immigrated to the United States and worked in a small electrical shop, named Osterheld and Eickemeyer, in Yonkers, New York. He had a command of electrical theory, and in 1892 he published an important paper on magnetic hysteresis. The paper aroused the intense interest of Elihu Thomson at the General Electric Company. The result was that General Electric bought the small company and engaged Steinmetz (see Figure IV.2) as consultant. Later, Steinmetz became General Electric's chief scientist.

International agreement on electrical units began with the First International Electrical Congress held in Paris in 1881. Volt, ohm, and ampere were among the units adopted. In 1889 a second congress convened, again in Paris. There it was agreed to adopt the units joule, watt, and henry (for inductance). At a third congress in 1891 units gauss (for magnetic intensity) and weber (for magnetic flux) were adopted.

The Fourth International Electrical Congress was held in Chicago in 1893 with Helmholtz as presiding officer. For that meeting the American Institute of Electrical Engineers appointed A. E. Kennelly as chairman of the American Standards Committee. Kennelly had published a paper on practical aspects of the AC theory in which he made use of Heaviside's work as well as Maxwell's treatment of resonance. In early 1893 he adopted Heaviside's term "impedance" as well as the use of complex quantities with $j = \sqrt{-1}$ for the alternating-current circuit elements. Steinmetz was impressed by this approach and adopted it himself. In fact, the paper, "Complex quantities," which he presented at the Congress and which was later published in *Transactions of the AIEE*, helped usher in the general use of complex numbers in computations of circuit responses.

Figure IV.2 A photograph of Charles Steinmetz.

The Congress also set international standards for electrical measurements. For example, the international ohm was defined as the resistance of a specified column of mercury, and the international ampere was defined by the amount of electrolytic deposition. (It also defined the international volt by decreeing the electromotive force of a Clark cell at 15° Celsius to be 1.434 international volts. Later, this definition was found to be inconsistent with the definitions of the ampere and the ohm, and it was abandoned in 1908.)

Eighteen hundred ninety-three can also be taken as the year when AC power systems gained supremacy over DC systems. It took some time, however, for standardization of frequencies and voltages to be achieved, even in a single region. The Westinghouse Company had successfully built an AC power plant in the Colorado mountains at Telluride that generated current at 60 hertz, and this frequency was eventually selected as standard for high-voltage distribution networks (see Figure IV.3). The European countries eventually settled on a 50-hertz standard for distribution networks. For transportation purposes, 25 hertz became the standard. (There were, of course, a host of technical and financial problems to be solved before the widespread use of AC power was feasible. One example is the need for a meter to indicate energy consumption, that is, a watt-hour meter. Such a device was invented in 1888 by O. B. Shallenberger.)

Steinmetz began a series of publications that made him the outstanding exponent of electrical engineering theory and practice in the United States. In January 1897 he published *Theory and Calculation of Alternating Current Phenomena*. This book is remarkable for its breadth of coverage. It consists of 30 chapters in 268 titled paragraphs on 401 pages, with two appendices totaling 91 pages. In the preface, he emphasizes, "...I have endeavored to make it as elementary as possible, and have therefore used only common algebra and trigonometry, practically excluding calculus, except for a section on the power aspects of transmission lines..."

Steinmetz followed this book by another, *Theoretical Elements of Electrical Engineering*, published in 1901. He also wrote 30 major papers and received some 200 patents. Steinmetz also played a key role in the creation of the General Electric Research Laboratory, which was organized in the fall of 1900.

IV.3 HEINRICH HERTZ AND THE DEMONSTRATION OF ELECTROMAGNETIC WAVES

Heinrich Hertz was born in Hamburg on February 22, 1857. After completing secondary school, he enrolled at the Dresden technical university (*Technische Hochschule*) to study engineering. He left after one semester for a required year of military service and then studied at the technical university in Munich for a year. Deciding that he preferred physics to engineering, he

Figure IV.3 Beginning in 1891 single-phase AC power, produced by a Westinghouse generator, was transmitted three miles from the Ames Plant in Telluride, Colorado, to the Gold King Mine. (Photograph courtesy of *Electrical West.*)

moved to the University of Berlin in 1878, where he studied under Hermann von Helmholtz and Gustav Kirchhoff.

Helmholtz and Kirchhoff, like almost all physicists on the Continent, held fast to the concept of action at a distance as formulated in Newton's theory of gravitation. The idea of "force field" as proposed by Faraday had no appeal. Of course, Hertz grew up in that atmosphere and was deeply influenced by the teachings of C. F. Gauss, W. E. Weber, and F. E. Newman.

Learning of the premature death of Maxwell in 1879, Helmholtz called for experimental verification or rejection of Maxwell's thesis of the existence of the displacement current and the induction of dielectric polarization by varying magnetic fields. He helped arrange for a prize of 100 ducats to be awarded by the Berlin Academy of Science for the best such proof or disproof. The final date of submission was set at March 1, 1881.

Helmholtz hoped that Hertz, who had been an outstanding student and had completed his doctorate in 1880, would take up this challenge. Indeed, Hertz's dissertation concerned electromagnetic induction in a rotating sphere, a topic very demanding mathematically. Hertz (see Figure IV.4) stayed on for three years as assistant to Helmholtz and reflected upon the task. Since the only known means of generating high-frequency oscillation was by means of spark discharges with an induction coil in the circuit—a problem treated by William Thomson in 1853—he could not readily visualize how one could achieve measurable effects.

Hertz continued, however, to think about the problem and to reflect on Maxwell's treatise. Having accepted in 1883 the position of lecturer (*Privatdozent*) in theoretical physics at the University of Kiel, he had good opportunity for that. He reformulated Maxwell's equations and arrived at very much the same duplex form as did Heaviside, whose results were appearing at about the same time in *The Electrician*. Though not yet convinced that Maxwell's bold hypothesis of a displacement current was the only way to proceed, Hertz considered it preferable to hypothesizing action at a distance.

In the spring of 1885 the technical university in Karlsruhe considered candidates for the professorship in physics. Helmholtz strongly recommended Hertz, who had just published a paper endorsing Maxwell's view. Though Hertz was only 28 years

Figure IV.4 A portrait of Heinrich Hertz (courtesy of the Burndy Library).

old, he won the appointment. The security of the full professor-ship permitted Hertz to plan for a family, and on July 13, 1886, he married the daughter of one of his faculty colleagues.

At his first visit to the laboratory, Hertz was impressed with the facilities that had been cared for by his well-known predeces-sor, Ferdinand Braun (1850–1918). When Hertz moved into the laboratory on March 29, 1885, he found among the equipment specially constructed spirals for the study of electrical discharg-es that appeared to provide sparks at very low power. This stim-ulated him to consider again whether he might prove Maxwell's hypothesis experimentally, and in the fall of the following year he started in earnest the planning of the experimental proce-dures.

Hertz realized at once that he needed to generate oscilla-tions at much higher frequencies. He was also concerned about the nature of the spark discharge, especially whether it could be represented appoximately as a damped sine wave. The first cir-cuit he devised to produce high-frequency oscillations consisted of an induction coil with a secondary coil connected to a spark gap, which had spheres for capacitive loading (see Figure IV.5).

Hertz calculated the resonant frequency of the circuit to be about 50 million cycles per second, which corresponds to a wave-length of 6 meters, a wavelength that was just short enough to

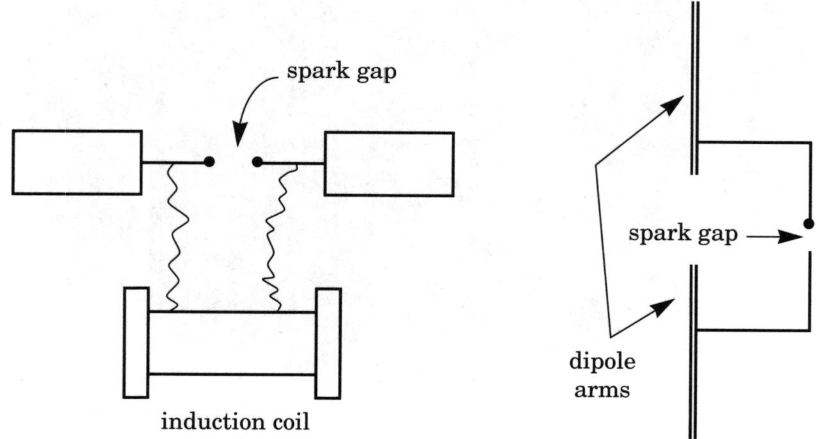

Figure IV.5 Hertz's circuits to produce and detect high-fre-quency oscillations.

permit the sort of laboratory investigation he intended. The receiver loop, which also contained a spark gap, could be adjusted for optimum response. In testing the setup, Hertz observed that the strong transmitter spark seemed to affect the receiver spark directly, and he found that by shielding the transmitter spark, he obtained greater consistency. (Hertz made notes of these observations, which later were recognized as involving the photoemission of electrons.) He used two coupled loops as receivers and explored with a probe for resonance.

Hertz then attacked the problem of demonstrating the presence of the displacement current in dielectrics. He achieved a higher frequency, about 100 million cycles per second (corresponding to a wavelength of 3 meters), so that sample sizes could be smaller. He used materials such as asphalt, paper (in the form of a pile of books), sandstone, wood, and petroleum, and, employing a circular detector ring that could be rotated, he was able to relate dielectric properties to the response of the detector spark gap.

With the higher frequencies, Hertz then demonstrated standing waves in space. By exploring with the detector spark gap, he could determine the locations of nodes and maxima and thus corroborate his earlier calculation of the wavelength. (By the relation, speed of propagation equals wavelength times frequency, he obtained a value for speed of the electromagnetic waves approximately equal to the speed of light. Measurements using spark gaps as indicators were necessarily rather crude.)

While carrying out some of these experiments, Hertz received a visit in April 1887 from Wilhelm von Bezold. In about 1870 von Bezold had conducted experiments with electric discharges, studying in particular the so-called Lichtenberg figures of discharge patterns. Apparently, von Bezold—like Oliver Lodge in England—had come close to recognizing wave propagation along wires.

Hertz was eventually able to raise the frequencies to about 500 million cycles per second (corresponding to a wavelength of 60 centimeters), and with these waves he was able to demonstrate reflection and focusing. He constructed parabolic cylinder reflectors and checked the polarization of waves (where the orientation of the wave is defined to be that of the electric field vector) by means of wire gratings that could be rotated.

Hertz gave a detailed account of these experiments in the book *Untersuchungen über die Ausbreitung der elektrischen Kraft* published in 1892. An English translation, entitled *Electric Waves*, appeared the following year. With his results Hertz believed that he had met the challenge of Helmholtz and the Berlin Academy of Sciences. In the preface to the book he noted, "The object of these experiments was to test the fundamental hypothesis of the Faraday-Maxwell theory and the result of the experiments is to confirm the fundamental hypothesis of the theory."

The book and earlier journal publications (especially those in Wiedemann's *Annalen*) attracted international attention and his experiments were repeated in many countries. In 1888 Oliver Lodge published a brief paper in *The Electrician* on "Measurement of electromagnetic wave length" to which he added the following:

> Since writing the above I have seen in the current July number of Wiedemann's Annalen an article by H. Hertz, wherein he establishes the existence and measures the wave length of aether waves excited by the coil discharges; converting them to stationary waves, not by reflexion of pulses transmitted along a wire and reflected from its free end, as I have done, but by reflexion of waves in free space at the surface of the conducting wall!

Hertz continued his investigations. He evaluated, from Maxwell's equations, the electric field lines produced by an infinitesimal vertical dipole at quarter-periods of the frequency cycle. He also computed the energy loss by radiation into space, but being aware of the space limitations in his laboratory and of the interferences resulting from reflections from ceiling and walls, he realized that exact experimental verification of the field values would not be possible in the laboratory. He also examined propagation along coaxial conductors and parallel wires, as well as the skin effect in metal conductors.

In 1889 he was appointed professor of physics at the University of Bonn, successor to Rudolf J. E. Clausius. Unfortunately, illness interfered with his work and caused his untimely death in 1894 at the age of 36. His book on a new approach to mechanics was completed by his assistant Philipp Lenard (who gained fame through his invention of the "Lenard window" in the study of cathode rays).

Hertz's was a short, but immensely influential, life. His work convinced many Continental physicists and engineers of the validity of Maxwell's theory, and the acceptance of Maxwell's theory came to have a profound effect on both physics and electrical engineering. And Hertz's work led, quite directly, to the exploitation of electromagnetic waves for communication, which is the subject of the next section.

IV.4 WIRELESS TELEGRAPHY

After the publication of Hertz's demonstration of electromagnetic waves in free space, his experiments were repeated in many countries. A major advance was the invention of a detector more effective than the spark-gap detector Hertz used. In 1890 in Paris, Edouard Branly (1846–1940) invented what he called a "Radiocompacteur," a glass tube containing metal filings. An electromagnetic wave caused the filings to cohere—readily detectable because of the increased electrical conductivity of the tube—and mechanical jarring restored the tube to its prior state. (Branly may have been the first to use the term "radio" for electromagnetic waves.) Improvements made in England by Oliver Lodge (1857–1940), who called the device a "coherer," led to its widespread use. A professor at the University of Bologna, August Righi (1850–1920), devised an extremely sensitive form of a spark-gap detector and experimented with very short waves, from 20 centimeters down to 2.5 centimeters.

Among the auditors of Righi's lectures was the young Guglielmo Marconi (1874–1937). Marconi became fascinated by electromagnetic waves and wondered why the possibility of their use for distant communication was not under discussion. He learned that in an 1892 magazine article William Crookes (1832–1919) had suggested the possibility, but wrote that more powerful and better controlled sources of high-frequency waves and more delicate receivers were needed. This impressed Marconi and during the winter of 1894–95 he experimented in the attic of his parents' villa. The following spring he moved his spark-gap transmitter equipment outside and used an elevated metal plate as a kind of antenna, connecting the other end securely to ground. With this arrangement and an improved coherer, he achieved reception over distances of 1 to 2 kilometers. Though he had little background in physics or engineering, his exceptional drive and perseverance made possible these advances.

Having achieved signal reception at considerable distances, Marconi tried to interest the Italian Ministry of Post and Telegraph. When he met with no success there, he turned his attention to England. His mother, who was Irish, had taught him English and was well connected in Britain, being a member of the family that owned the famous Jameson distillery. So on February 2, 1896, Marconi and his mother left Italy for England. He at once applied for a patent, "Improvements in Transmitting Electrical Impulses and Signals and in Apparatus Therefor," which was granted that same year.

A meeting was arranged with William H. Preece, the chief engineer of the British General Post Office, who had experimented himself with induction between wires. Preece expressed interest in this new approach and arranged for a demonstration by Marconi in August 1896. When this demonstration turned out well, another was arranged for a distance of almost 3 kilometers, and on the last day of 1896 Preece promised at a public lecture that the British post office would spare no expense in developing this epochal invention.

In March 1897 Marconi gave a demonstration of his system over a distance of more than 5 kilometers at the Bristol Channel. To this demonstration, Preece invited Adolf Slaby (1849–1913) of the technical university in Berlin-Charlottenburg, who was science and technology advisor to the German emperor. Marconi was displeased by Slaby's presence, because he was aware of German competitors.

One of the German competitors was Karl Ferdinand Braun (1850–1918), who had studied at the University of Berlin with Herman von Helmholtz. After a brief stay at the University of Würzburg as laboratory assistant, where he discovered the rectifier effect in metal sulfides, Braun moved first to the technical university of Karlsruhe for two years (just before Heinrich Hertz's years at Karlsruhe) and then to the University of Tübingen, where he stayed from 1885 to 1895. There he developed an electrostatic voltmeter and investigated thermoelectricity. In

1895 he accepted the professorship in physics at the University of Strassburg (then in Germany) as successor to F. W. Kohlrausch (1840–1910), who became director of the German National Physical Laboratory. Braun brought with him to Strassburg his former student Jonathan Zenneck (1871–1959) as his assistant.

In the modern laboratories of the University of Strassburg, Braun concentrated his efforts on cathode rays. These had first been studied by Julius Plücker in Bonn, where Heinrich Geissler (1814–1879) had constructed the most effective glow tubes. Wilhelm Roentgen's announcement of the discovery of x-rays on New Year's Day 1896 increased Braun's interest in cathode rays. He found that by the use of a magnetic field he could deflect the ray vertically. This gave him the idea of using the deflection to trace the time variation of alternating currents by means of a rotating mirror. He began these experiments in the summer of 1896, and the following February he reported his invention of the cathode ray oscilloscope, now indispensable in laboratories and the basic element of electronic television. Braun's assistant Zenneck later added the mechanism for horizontal deflection. Braun never patented this invention, nor did Hertz and Roentgen ever apply for patents on their novel contributions. These people apparently felt that these basic contributions should be freely available and not classified as private property.

Braun was invited to give a presentation of his instrument at a scheduled meeting of the British Association for the Advancement of Science. The meeting, quite exceptionally, was held outside the British Isles, in Toronto, Canada, in August 1897. Braun made the voyage, and on his return he became very much interested in wireless telegraphy. Earlier experimentation gave him the idea for a new transmitter circuit: if one inserted an inductive coil antenna and ground and coupled it magnetically to another coil in the transmitter circuit, the transmitter circuit would be separated from the antenna circuit and more power would be available for radiation. After successfully demonstrating the circuit, he filed a patent application on October 14, 1898, and founded a company called *Funkentelegraphic GmbH, Köln* (Funkentelegraphic Ltd., Cologne). Strong financial support from the Hamburg business community led to the formation on July 7, 1899, of an encompassing syndicate called Telebraun.

Adolf Slaby, who witnessed Marconi's demonstration in March 1897, established contact with AEG (*Allgemeine Elektrizitäts Gesellschaft*) in Berlin, where his former student Georg Wilhelm Alexander Hans Graf von Arco (1869–1940) was in charge of the radiotelegraphy laboratory. Slaby developed competitive circuits and joined AEG in December 1900, which thus became a serious competitor to Telebraun. In response, Braun allied himself with Siemens, and the *Braun-Siemens Gesellschaft* was formed in July 1901.

In the meantime Marconi was moving ahead with commercial exploitation of his inventions. On July 13, 1897, he applied for a U.S. patent, and on July 20, 1897, he organized the Wireless Telegraph and Signal Company, Ltd. with funds provided by family and friends and with his cousin Henry Jameson Davis as chairman of the Board. The company acquired some of Oliver Lodge's patents (which allowed some crude tuning of the antenna circuit) and the services of Lodge as consultant.

Marconi achieved public acclaim when his first permanent transmitting and receiving installation on the Isle of Wight was visited in June 1898 by Lord Kelvin (William Thomson), who sent telegrams to Preece and to George G. Stokes (1819–1903). Equally newsworthy was Queen Victoria's request that Marconi establish stations able to communicate with the royal yacht. In March 1899 Marconi succeeded in transmitting across the English Channel to France, and that summer he provided wireless service to the Royal Navy during maneuvers.

In 1900 the company changed its name to the Marconi Wireless Telegraph Company. In the same year Marconi filed for a patent on a system of tuning (employing for the first time a tuning dial). This was the famous British patent number 7777, which was the most important of the many patents held by Marconi's company that secured its domination of wireless telegraphy in many countries.

The early decades of radio were marked by numerous lawsuits concerning the validity and infringement of patents. The public disclosure required in patenting probably worked to the detriment of some inventors. It appears, for example, that almost all the patent disclosures of Braun were bit by bit incorporated into the operating circuits of competitors. But public disclosure served to advance the science and techniques of radio.

The University of Strassburg was the first to establish an institute for radio science, and some of the inventors, notably Lodge and Braun, were concerned to advance radio science. (In the 1890s Lodge wrote several articles for *The Electrician* on the scientific basis of radio, and in 1901 Braun published an influential paper entitled "On rational transmitter circuits in wireless telegraphy.")

The next major objective of Marconi was now to cross the Atlantic Ocean with radiotelegraphy (see Figure IV.6). He visited the United States in 1900 and, upon return in the fall, started construction of a large antenna set at Poldhu in Cornwall, England. In December of the following year at St. John's in Newfoundland, he succeeded in receiving signals from the Poldhu station.

In 1902 when Marconi was sailing to England on the U.S. liner *Philadelphia*, he found that he could receive clear signals from Poldhu at distances as great as 1700 miles, but that there were great variations in intensity from day to night. This experience raised two basic questions: How could electric waves be received over such large distances if their propagation is rectilinear as demonstrated by Hertz? and What accounts for the variation in reception from day to night? In 1902 Arthur Edward Kennelly (1861–1934) and Oliver Heaviside, working independently, hypothesized the existence of a conductive layer of ions high in the atmosphere that reflected the electric waves back to earth. Since the height and density of the conductive layer (called initially the Heaviside-Kennelly layer, later the ionosphere) might well be affected by the sun's radiation and thus the time of the day on earth, this hypothesis could explain both the long distance of transmission and the variability that Marconi experienced. The supposition of a layer of ions (actually several layers) was subsequently verified, beginning with the work in 1926 by Gregory Breit (1899–1981) and Merle Tuve (1901–1982), who were able to record the reflections of pulses of electric waves propagated vertically and thus measure the height of the Heaviside-Kennelly layer.

Over the years the effectiveness of radio telegraphy was gradually improved in a number of ways: the move from spark generation of electric waves to the generation of continuous waves, the attainment of greater transmitting power, the im-

Figure IV.6 A photograph of Guglielmo Marconi (left) and George Kemp taken shortly after the successful transatlantic test of 1901.

provement of radio receivers, the development of practical means of measuring wavelength or frequency of electrical waves, and the ability to control partially the direction of radiation.

Marconi was using essentially Hertz's spark-gap technique to generate short waves. Already in 1900, William DuBois Duddell (1872–1917), an English telegraph engineer, attempted to use a DC electric arc in a resonant circuit to obtain sustained oscillations. With his so-called "singing arc," he succeeded with frequencies up to 10 kilohertz (that is, 10,000 cycles per second) (When, in 1922, I took a course on electric waves at the technical university in Vienna, we were given a demonstration of the singing arc.)

In Denmark in 1902, Valdemar Poulsen devised a way to use an electric arc to produce undamped, continuous waves at about 1 MHz (that is, 1 million hertz). He obtained a Danish patent and later an English patent, but because of the near global dominance of the original Marconi patents and the fact that in 1906 E. F. W. Alexanderson, working with the General Electric Company in the United States, submitted a high-frequency electric generator to final tests, Poulsen's chances of success appeared slim. However, the U.S. Navy, which had established wireless stations on the West Coast by 1904, was very much interested in high-power transmitters and were to become principal supporters of Poulsen's system.

Cyril Elwell, a graduate of Stanford University, learned of Poulsen's work, went to Copenhagen in May 1909, and with financial support from California friends, arranged for Poulsen to move with an assistant to Stanford. In October 1909 Poulsen organized the Poulsen Wireless Telephone and Telegraph Company, which established telegraph stations furnished with Poulsen arc transmitters. Because of the greater power of the arc transmitters and the continuous wave operation that permitted telephony as well as telegraphy, the company grew. In 1910 it became the Federal Telegraph Company, and by 1912 the company's transmitters were practically standard at U.S. Navy stations.

The principal alternative to the Poulsen arc transmitter for continuous oscillations was the alternating-current generator invented in 1901 by Reginald Fessenden (1866–1932) at the National Electric Signaling Company in Boston. He convinced

Charles Steinmetz at the General Electric Company to build a higher-frequency generator. (See Figure IV.7.) Steinmetz, after designing a generator capable of 10 kHz, turned the effort over to a young Swedish engineer, Ernst F. W. Alexanderson (1878–1975), who had studied with Slaby and had joined GE in 1902. In 1906 Alexanderson achieved 80 kHz—the generator turned at a rate of 20,000 rpm—with a power output of 1 kilowatt (kW). Fessenden used this machine at his radio station in Brant Rock, Massachusetts, to transmit a Christmas program over the air in December 1906.

The means for detection of wireless telegraph signals also progressed. Braun's discovery of the rectifying effect in crystals was exploited in the crystal detector, which became the detector used in most amateur receivers the world over. The coherer was replaced by detectors using the "Fleming valve" or diode vacuum tube, invented in 1904 by John Ambrose Fleming (1849–1945), who made a useful device of the phenomenon Thomas Edison had discovered years earlier (of the flow of current in a tube from heated cathode to anode; see Figure IV.8).

As radio engineering advanced and as governments took on regulatory functions, it became increasingly important to have practical means of measuring wavelength or frequency of radio waves. In 1907 a wavemeter was produced by Siemens-Halske using a tuned circuit with variable components that could be calibrated directly in wavelength. In the United States, J. V. L. Hogan designed a wavemeter that was standardized in 1911 by the National Bureau of Standards.

Finally, gaining some control over the direction of radiation became imperative, partly because interference between different transmitters became more and more troublesome, partly because greater distances could be reached if the radiated power could be concentrated. As early as 1900 Zenneck experimented with placing transmitting towers in a line to enhance the radiation in the direction of the line. In 1906 Braun published accounts of his experiments with phased-delayed transmission (for achieving directionality), and Marconi also developed horizontal directional antennas, principally for transatlantic service.

The development of the art and science of radio occurred on an international stage. This is accounted for partly through the exchange of information across borders—made possible by scien-

Figure IV.7 A photograph of the twin 200-kilowatt alternators at the Marconi Carnarvon station (courtesy of General Electric Research and Development Center).

Figure IV.8 A photograph of an early Fleming valve, circa 1906 (courtesy of the Smithsonian Institution).

tific publications, patent disclosures, and other writings—and partly because of the international aspirations of some companies. But governments also became involved. The British government protected the interests of the Marconi company, and in 1903 Emperor Wilhelm II demanded that the rivalry between

Siemens-Halske and AEG be turned into cooperation, and this resulted in the creation of Telefunken as the German counterpart to the British Marconi Wireless Telegraph Company. After World War I, concern that radio in the United States would be dominated by a foreign company (Marconi) led to the formation of the Radio Corporation of America.

Another expression of the international character of radio science was the award in 1909 of the Nobel Prize in Physics to Marconi and Braun jointly. On December 10, 1909, Hans Hildebrand, president of the Swedish Academy of Sciences, introduced the two award winners with the statement:

> A man was needed who was able to grasp the potentialities of the enterprise and who could overcome all the various difficulties that stood in the way of the practical realization of the idea. The carrying out of this great task was reserved for Guglielmo Marconi.... The development of a great invention seldom occurs through one individual man, and many forces have contributed to the remarkable results now achieved. Marconi's original system had its weak points.... It is due above all to the inspired work of Professor Braun that this unsatisfactory state of affairs was overcome.

IV.5   Electrical Technology at the Turn of the Century

The last decade of the nineteenth and the first decade of the twentieth century saw two fundamental technological transformations: world-spanning, wireless telecommunications became a reality, and the provision of electric power became a mature industry. The former, which was the subject of the last section, can be seen as growing out of Hertz's demonstration of the existence of electromagnetic waves. The latter, one may argue, grew out of Edison's invention of a practical incandescent light bulb.

Soon after his invention of the practical light bulb, Edison recognized the need for large-scale generators of electric power plants and for extensive systems for power distribution to meet the growing demand for electric power. Direct current, which he favored, was theoretically simpler, but impractical for a large-scale distribution system. In the United States it was George

Westinghouse—and engineers, such as Nikola Tesla, working for Westinghouse—who saw the advantages of alternating current and who worked to bring about high-voltage systems. One of Westinghouse's first large installations was at Telluride, Colorado, in 1891. There, at an elevation of 9000 feet, a waterwheel generator of 100 horsepower generated single-phase alternating current at 3000 volts, which was transmitted 3.6 kilometers before being used for mining operations. In Europe, S. Z. de Ferranti built an AC system supplying part of London in the late 1880s, and the Swiss firm of Brown-Boveri completed an AC system in Germany in 1891.

Other installations followed with higher voltages and greater power. At every step of the way, engineering challenges had to be met. For example, with voltages above 20,000, corona discharges were observed, accompanied by appreciable energy losses. This became a subject of special interest for Harris J. Ryan, a teacher at Cornell University. In 1905 Ryan became head of the Department of Electrical Engineering at the recently opened Stanford University at Palo Alto, California, and in 1913 he established there a high-voltage laboratory. Progress was astonishing: in 1900, operating voltages were as high as 50,000 volts; by 1910, they had progressed to 110,000 volts; and in 1920, the major transmission lines, both in Europe and in the United States, operated at 220,000 volts.

In the mid-1890s at Niagara Falls, New York, a large project to generate, distribute, and use electricity was carried out. This project began to produce power in 1895, when it provided electricity locally, notably to the Pittsburgh Reduction Company (later renamed Alcoa). The following year, three-phase AC power at 11,000 volts began to be supplied to Buffalo, 20 miles away. The Niagara system was a synthesis of European and American advances in large hydraulic turbine design (see Figure IV.9), and it used the three-phase power that is now fairly standard.

One application where three-phase power is not standard is electric traction, which usually employs single-phase, low-frequency power to suit the motors used in electric trolleys and trains. (DC power was also frequently used for trolleys.) Electric trolleys for urban areas (see Figure IV.10) became extremely important in the 1890s, and interurban trolleys and the electrification of railroad lines soon followed.

Figure IV.9 The dynamo room at the Niagara Falls power station in 1905 (courtesy of the Smithsonian Institution).

Figure IV.10 An illustration of the electric trolley in Pittsburgh.

Until the end of the nineteenth century the most important prime mover was the steam engine. Toward the end of the century two formidable competitors appeared: the internal combustion engine and the steam turbine. It was the latter which soon had a large impact on the electric power industry. In generating electric power, the steam engine required large flywheels to convert reciprocating motion into smooth rotational motion. This was not true of the steam turbine, developed by Carl G. P. de Laval in Sweden and Charles Algernon Parsons in England. Parsons, who organized his own works in Heaton, Newcastle upon Tyne, achieved great success. He became associated with Westinghouse, and the first steam turbine in the United States was installed in 1901 at Hartford, Connecticut. Further development by Charles Gordon Curtis led to patents which Curtis sold to the General Electric Company. By 1904 both Westinghouse and General Electric were receiving many orders for turbine generators.

Another area in which electrical technology made great advances in the years around 1900 was long-distance wire communications. By 1887 Heaviside recognized, from his mathematical analysis of signal propagation along transmission lines, that signal preservation depends crucially upon proper relations between the transmission line parameters as expressed by the criterion $r/1 = g/c$. In almost all practical cases, this demands increasing greatly the inductance per unit length of line. Heaviside, however, suggested no practical means of satisfying this condition. When in 1890 Bell Telephone Company hired John Stone Stone (1869–1943) from Johns Hopkins University, he immediately became interested in applying Heaviside's theory to long-distance telephone circuits. Stone had earlier solicited Heaviside's help directly. In 1897 George Ashley Campbell (1870–1954) from Harvard University became assistant to Stone, and both concentrated on applying Heaviside's theory to the junction of open-wire transmission line and cable by the consideration of energy reflections. Campbell, indeed, used his loading coil selection in his 1901 doctoral dissertation for Harvard University, and he proposed a design of loading coils for Bell telephone lines.

Michael Idorsky Pupin (1858–1935) at Columbia University had also read Heaviside's exposition of the need for inductive loading of telephone transmission lines and had even given praise to Heaviside in a paper published in 1900 in the *Transactions of the AIEE*. Nevertheless, Pupin claimed credit for the invention of the loading coil, and he persuaded the Bell Telephone Company to pay him handsome royalties. (Later the patent claims of both Campbell and Pupin were rejected because of Heaviside's prior publication.) Both Campbell and Pupin suggested that inductance be increased not uniformly along the transmission line, but at regular intervals. Inductive loading proved effective in telephone transmission.

Continuous, as contrasted with discrete loading, was employed in submarine cables. The first inductively loaded submarine cable was laid across the Baltic Sea in 1902 and 1903 by Carl Emil Krarup. He used soft iron wire spiraled around the copper-core conductor. Substantial improvement came with the use of a powdered nickel-iron alloy (permalloy) discovered by G. W. Elmen of Western Electric in 1915.

The years around 1900 saw advances in the profession of electrical engineering. As we have already seen, the American Institute of Electrical Engineers was formed in 1884. It was preceded by the American Society of Civil Engineers (1852), the American Institute of Mining Engineers (1871), and the American Society of Mechanical Engineers (1880). The increasing importance of the relatively new professions of engineering led in 1893 to the formation of the American Society for the Promotion of Engineering Education. The first meeting of the Society, held at Polytechnic Institute of Brooklyn, was attended by 70 members who agreed to invite the professional engineering societies to collaborate with the new society in devising appropriate curricula for the different courses of study. In 1902 Samuel Sheldon of the Polytechnic Institute presented a paper at the 19th annual convention of AIEE, giving a comparative table of principal topics with maximum and minimum hours-per-week instruction in a large number of institutions; Sheldon pointed out the rather substantial differences among institutions.

Six years later Charles Steinmetz took up the matter of electrical engineering education at a special meeting of AIEE. He pointed out that there was a tendency to demand a technical college training even for administrative and commercial positions. He contended, though, that the amount of material crowded into a four-year course was beyond that which even the better students could possibly digest. He argued instead for EE education aimed at a thorough understanding of the fundamentals of the discipline. Achieving this, he said, would require a higher grade of teachers than were being attracted by the meager remuneration then given. Steinmetz, who had received his education in Europe and who had long been teaching at Union College in Schenectady, said, "I believe that in the teaching of allied sciences, our colleges and schools are still greatly inferior to those abroad; the result is very marked in the product of the colleges in the inferiority of the general education possessed by our graduates."

One reason that both Sheldon and Steinmetz were concerned with high educational standards for electrical engineers is that it appeared that the AIEE was becoming dominated by businessmen and managers rather than scientifically trained engineers. This was partly due to the vigorous growth of the electric power industry in those years (roughly doubling in size every five years).

Another consequence of the vigor of the electric power industry was that some AIEE members felt that other areas of electrical engineering were being slighted. In response, committees for technical specialties began to be formed. The first two, formed in 1905, were the Telegraph and Telephone Committee and the High Tension Transmission Committee. Eleven other technical committees were established by 1913. In one technical area, wireless telegraphy, the AIEE moved so slowly that an independent society was formed and won the allegiance of most practitioners of the young art of wireless communication.

Already in 1907 John Stone established the Society of Wireless Telegraph Engineers (SWTE) in Boston. The following year Robert Marriott founded the Wireless Institute in New York. In early 1912 Marriott and John V. L. Hogan of SWTE met with Alfred Goldsmith of City College of New York to discuss the lack of recognition of the new mode of communication. This discussion led to a much larger meeting on May 13, 1912, in Fayerweather Hall of Columbia University, and there the Institute of Radio Engineers (IRE) was formed. (See Figure IV.11.) Marriott was the first president, Fritz Lowenstein the vice president, and Alfred Goldsmith the editor of the society's *Proceedings*. The jour-

Figure IV.11 A photograph of the IRE banquet at Luchow's Restaurant in New York City on April 24, 1915. Among those present were E. F. W. Alexanderson, Ferdinand Braun, Lee de Forest, John Stone Stone, and Nikola Tesla.

nal did much to establish the reputation of the new society, and its membership grew steadily. A proposal to add "American" to the IRE name was rejected on the grounds that the society, just as the radio art, should be international in scope. The journal did, indeed, gain an international readership, and the Institute attracted increasingly many members from outside the United States. A Canadian Section of the IRE was established in 1925, and a Brazilian Section in 1939.

IV.6 ELECTRICAL SCIENCE AT THE TURN OF THE CENTURY

In the nineteenth century a common view among physicists was that a subtle substance, called the ether, pervaded all of space and permitted heat and light to propagate across seemingly empty regions. The discovery of electromagnetic phenomena, which exhibited several forms of action at a distance, made this hypothesis even more acceptable, and it was endorsed by leading physicists such as Oliver Lodge and William Thomson. The recognition, by Maxwell and Hertz, of light as an electromagnetic phenomenon, together with the theoretical framework of Maxwell's theory, seemed to characterize the ether as the electromagnetic medium.

There seemed no reason to suppose that the ether was at rest with respect to the earth, so attempts were made to measure the movement of the earth relative to the ether. The French physicists Armand Fizeau and Léon Foucault had succeeded in measuring the velocity of light in the laboratory, and physicists hoped that motion of a laboratory relative to the ether might be detectable by extremely exact measurements of the same sort.

Albert Abraham Michelson (1852–1931) was brought by his Polish parents to the United States at age 2. His youthful ambition was to be a midshipman in the U.S. Navy, and he won admission to the Naval Academy. Upon graduation in 1873, however, he was assigned to teaching physics. He soon became interested in optics, and making skillful use of simple experimental equipment, he measured the velocity of light in 1878 with remarkable accuracy. His meticulous repetition of the measurement with better equipment and higher accuracy gained him the attention of the American physics community. In 1880 Michelson went to Europe to study at the École Polytechnique in Paris and the universities of Berlin and Heidelberg.

Still a naval officer, he then concentrated on means to determine the effect of the hypothetical ether upon the propagation of light. For this purpose he invented in 1881 the interferometer (a device that allows precise determination of displacements and wavelengths by a system of mirrors and glass plates causing a beam of light to travel two different optical paths), and with it he made an attempt to verify or disprove the "ether drift" (that is, the motion of the earth relative to the ether). He showed that the observed shift of interference fringes was about 1/40th of that expected if the ether was stationary, and he concluded, "the hypothesis of a stationary ether is thus shown to be incorrect." Because of the importance of this result, Michelson, who in 1883 was appointed professor of physics at the newly organized Case Institute for Applied Science in Cleveland, joined with Edward W. Morley to refine the measurements.

The general reaction to Michelson's result was disbelief, but two people—the Irish physicist George Francis Fitzgerald (1851–1901) and the Dutch physicist Hendrik Lorentz (1853–1928)—independently proposed that the ether hypothesis could be retained if one supposed that the interferometer arms contracted in the direction of the earth's motion through the ether. Lorentz grounded this supposition in a theory of matter as composed of charged particles, according to which the forces determining the size of bodies are propagated by the ether.

In 1905 Albert Einstein showed that the Michelson-Morley result could be understood in a different way. He rigorously deduced the consequences of two assumptions: that the laws of physics take the same form in all inertial reference frames (the postulate of relativity) and that in a given inertial reference frame, the speed of light is the same regardless of the speed of the body emitting the light (the postulate of the constancy of the speed of light). This was, of course, Einstein's special theory of relativity. It turned out that Maxwell's equations were relativistically invariant, but that Newton's mechanics had to be reformulated. Einstein showed in particular that a moving object appears to contract in the direction of the motion, the amount of contraction being exactly what Lorentz had earlier proposed. Einstein's theory, which made no postulate of an ether, eventually had the effect of eliminating that concept from physical theory.

Even as electricity was being supplied to homes, businesses, and factories in the 1890s, engineers and physicists had no clear idea of the nature of electricity, even whether it was in essence a wave or a particle. On the one hand, the wave theory of light had early in the nineteenth century won ascendancy over particle theories, and Maxwell's theory depicted light as a wave phenomenon. On the other hand, controlled electric discharges (called cathode rays) in evacuated glass containers, as investigated by William Crookes and others, suggested that the discharge was a stream of charged particles.

It was J. J. Thomson (1856–1940) at the Cavendish Laboratory at the University of Cambridge who first convincingly demonstrated the particulate nature of cathode rays. In 1897 he arranged, in an evacuated tube, a hot filament to emit electric charges, an accelerating anode, and, what was novel, perpendicular electric and magnetic fields in the space beyond the anode. By adjusting the strength of the electric field so that its effect is exactly balanced by that of the magnetic field (shown by the lack of deflection of the cathode ray), Thomson (see Figure IV.12) was able to calculate the ratio of charge to mass—denoted e/m—of the cathode ray particles.

This ratio was, quite surprisingly, almost 800 times that of ionized hydrogen, and Thomson surmised that the cathode ray particle (which soon came to be called the electron) was a sub-atomic particle and a constituent of all atoms. Working with his student C. T. R. Wilson, who had invented what came to be called the cloud chamber, Thomson in 1899 made an approximate measurement of the charge of the electron. With this value, Thomson calculated the electron mass, which was indeed less than a thousandth that of hydrogen, the lightest element.

The last decade of the nineteenth century was one of the most eventful in physics. Besides the work of J. J. Thomson and others on cathode rays, there was the discovery by Wilhelm Roentgen, announced on January 4, 1896, of x-rays. In the following month, Henri Becquerel reported to the French Academy of Sciences on the discovery of the radioactivity of uranium, and his associates, Pierre and Marie Curie, very soon discovered oth-

Figure IV.12 A portrait of J. J. Thomson.

er radioactive substances. The first decade of the new century was equally eventful.

A scientist who was then coming into prominence was Robert Andrews Millikan (1868–1953). His mentor at Columbia University, Michael Pupin, insisted that he should spend some time in Europe, so in 1895 Millikan, after completing his Ph.D.,

Figure IV.13 A photograph of Robert Andrews Millikan.

took a course from Max Planck in Berlin and did research with Hermann Nernst in Göttingen. Before leaving for Europe, Millikan had paid a visit to the newly opened University of Chicago, where he met Albert Michelson. He must have made a favorable impression, because in the summer of 1896 he received a cable from Michelson offering him an assistantship at the University of Chicago.

Though J. J. Thomson and others had approximate determinations of the charge of the electron, an exact measurement of this fundamental constant was widely desired. Millikan (see Figure IV.13) started with a method that had been introduced by the physicist H. A. Wilson: one measured the rate at which a charged cloud of water droplets fell in the absence and in the presence of an electric field. He found that he could work with single droplets, and thus obtain better measurements. He presented preliminary results in September 1909 at the meeting of the British Association for the Advancement of Science held in Winnipeg. On the train back to Chicago, it suddenly occurred to him that atomized oil droplets should prove better than water droplets, since oil was not nearly so subject to evaporation. In the fall of 1910 Millikan attracted worldwide attention when he published his measurement of the charge of the electron as 1.6022×10^{-19} coulomb.

Work going on in another branch of physics came to have great importance for electrical engineering, though few would have predicted that at the time. This work concerned heat radiation from hot bodies. Finding quantitative laws for this emission appeared extremely difficult because of the many factors that might influence intensity and wavelength of the energy radiated, such as temperature, nature of the material, and character of the surface. Physicists sought something that could serve as a standard, for which they introduced the ideal "blackbody" radiator, approximated experimentally by a small hole in a large furnace enclosure. Among those who investigated the dependence on temperature of the energy emitted were Joseph Stefan, Gustav Kirchhoff, and Wilhelm Wien, and several quantitative laws were proposed. According to what became known as the Stefan-Boltzmann law, the total energy should vary with the fourth power of the absolute temperature. The so-called Wien displacement law predicts that the shorter the wavelength, the higher the maximum energy, but Wien was unable to find a mathematical relation for the dependence on temperature and wavelength except for short wavelengths. In England John Strutt (Lord Rayleigh) and James Jeans found an approximate relationship for the long wavelengths.

The person who succeeded in harmonizing and unifying these partial results—and who opened up a new realm of physics in doing so—was Max Karl Ernst Ludwig Planck (1858–1947). Planck studied at the universities of Munich and Berlin, and in 1889 he became a professor at Berlin, the successor to Kirchhoff. He was deeply troubled by the inability to find a formulation for the wavelength distribution of the energy emitted by a blackbody. In 1900 he finally succeeded, but only by abandoning a basic tenet of physics at the time—that the energy of an elementary oscillator can vary continuously. Planck hypothesized, in contrast, that the oscillator takes on only particular energy values, all integral multiples of the smallest value, and he called each energy unit a quantum of energy. At a meeting of the German physics society on December 14, 1900, Planck presented his law of blackbody radiation, showing that it was consistent with the Stefan-Boltzmann law and that it reduced to the Wien law for short wavelengths and the Rayleigh-Jeans law for long wavelengths.

Among those deeply impressed by Planck's work was Albert Einstein (1879–1955). Einstein related it to another surprising finding—the experimental finding by Hertz that ultraviolet light could charge metal surfaces. In 1902 Philipp Edward Anton Lenard (1862–1947) found that the maximum kinetic energy of the electrons that the ultraviolet light causes to be emitted depends only on the frequency of the light, not on the intensity. From this, Einstein inferred that the light energy interacts only as discrete units.

When he heard of this hypothesis, Robert Millikan had grave doubts. In October 1912 he set out to get accurate experimental data to confirm or disconfirm Einstein's hypothesis. By 1915 Millikan had confirmed Einstein's equation of the photo-effect in detail, but remained doubtful of the underlying quantum hypothesis. Nevertheless his work was a stimulus to the development of a wide-ranging and extraordinarily successful quantum theory that later became important in electrical engineering.

IV.7 RADIO

Radio, or wireless as it was usually called, became an important business in the first decade of the century. This was, of course, radio telegraphy rather than telephony, and it was used mainly as a means of communication with ships at sea. The British Marconi company, along with its affiliates in several countries, enjoyed a dominant position in this business. During World War I the technology assumed great significance for military and dip-

lomatic communications, the more so as many submarine tele-
graph cables were intentionally severed. The American Marconi
Company built a powerful transmitting station in New Bruns-
wick, New Jersey. The station, which was taken over by the U.S.
Navy, used a 200-kW Alexanderson alternator (see Figure
IV.14).

Both the art and science of radio made steady advance. An important contribution to the science stemmed from work George

Figure IV.14 A 1922 photograph of Ernst Alexanderson with
one of the high-frequency alternators he designed
(courtesy of General Electric Research and Development Center).

A. Campbell (1870–1954) was doing in 1903 on long-distance telephone transmission. Using laboratory simulations of different sorts of cable, he found that he could not only cut out harmonics in generator currents, but by combining inductances and capacitances he could cut out all frequencies below a certain limit, thus producing what could be called a wave filter. Initially Campbell did not get an encouraging response from the patent office, so it was not until 1917 that his patent was issued. In the meantime, in 1915, a competing patent was issued to Karl Willy Wagner of the research group in the German post office. In 1922 Campbell published a comprehensive paper on wave filters (in the *Bell System Technical Journal*), which was followed by a paper by O. J. Zobel with specific design information.

This work is among the most ingenious and valuable of all contributions to communications technology. Wave filters permit the channeling of spoken information into specific frequency ranges, which extend from the low-frequency telephone circuits to the microwave and optical frequency ranges. Engineers can thus effectively use an enormous frequency spectrum without mutual interference.

Interference was a serious problem in the early days of radio broadcasting. Before there could be broadcasting, however, there had to be an effective generator of continuous electromagnetic waves (in contrast to the rapidly damped bursts of radiation produced by spark-gap transmitters). That generator was found in the electron tube.

In 1904 J. A. Fleming, with his invention of the diode, showed how the Edison effect—the electric current from a heated cathode in an evacuated tube—could be used to rectify current. Oliver Lodge used the diode as a detector of radio waves. In 1906 Lee de Forest found that by adding a third electrode to the vacuum tube, he could achieve amplification of reception. However, this triode vacuum tube, which de Forest called the audion, was neither reliable nor stable in operation. In 1911 Fritz Lowenstein showed that giving the third element a negative rather than a positive potential, as de Forest had done, improved stability greatly.

At the General Electric Company, E. F. W. Alexanderson procured an audion and aroused the interest of Irving Langmuir (1881–1957) in the device. Langmuir studied the potential distribution in the triode and space charge effects, leading him to the discovery of the famous $3/2$ power law for the current-voltage relations. Langmuir also discovered that filling the tube with an inert gas reduced filament evaporation. (This made "vacuum tube" a misnomer in some cases, but it was not until the 1920s that the term "electron tube" came into use.) Langmuir and others at General Electric developed a highly effective amplifier tube, the pliotron, in 1913, and the company entered the tube business.

In 1912 Edwin Howard Armstrong (1890–1954—see Figure IV.15) made use of the triode in a regenerative circuit, which gave amplification by positive feedback; others were devising similar circuits at about the same time. In about 1916 Armstrong made another invention, the superheterodyne receiver, in which the radio frequency signals were converted to a lower frequency prior to amplification and detection.

The Marconi Company obtained a dominant position in North America when it purchased the United Wireless Company (which had 70 shore stations) after winning a patent infringement suit in 1912. The dominance was enhanced by winning a similar suit against another large U.S. company, the National Electrical Signaling Company, which later went out of business. World War I made clear to military and government leaders the importance of radio, and when in 1919 the Marconi Company moved to acquire rights from General Electric to the Alexanderson alternator, the government feared complete domination of radio by a foreign company. At the urging of President Woodrow Wilson, General Electric purchased the American Marconi Company and set up a new company, the Radio Corporation of America (RCA). AT&T and Westinghouse, both of which had some stake in the radio business, soon entered the RCA group.

Thus the stage was set for the rapid establishment of radio broadcasting. Perhaps the first station with regularly scheduled and announced broadcasts was Pittsburgh's KDKA, owned by Westinghouse. KDKA's first broadcast (see Figure IV.16) was the announcement of the election returns on November 2, 1920, when Warren G. Harding and Calvin Coolidge easily outpolled James M. Cox and Franklin Roosevelt. By 1922 there were already over 500 broadcasting stations in operation, and by the end of the decade half of all American homes had radios.

Figure IV.15 This photograph shows Edwin Howard Armstrong, who was throughout his life fascinated with heights, atop the WJZ antenna in New York City in 1923.

Because radiotelephony by means of amplitude modulation (AM) was strongly affected by interferences of many kinds, E. H. Armstrong developed a system of frequency modulation (FM) as a better way of broadcasting. Armstrong tried without success to interest RCA in FM radio. RCA, which had a large stake in the business of AM broadcasting, was not interested in developing a competing broadcasting system. (An additional consideration was that FM stations would require greater bandwidths, and Armstrong hoped to use a part of the spectrum that RCA ex-

Figure IV.16 A photograph of the first broadcast from radio station KDKA, which consisted of reports from the Harding-Cox election returns (courtesy of Westinghouse Corporation).

pected would be given to television broadcasting, as soon as the latter became a reality.) Armstrong then embarked on the development of FM at his own expense, and in 1939 he began FM broadcasting from his own station (W2XMN) at Alpine, New Jersey.

The rapid development of radio engineering, which was only gradually reflected in the courses offered at schools of electrical engineering, contributed to a growing dissatisfaction with the existing EE curricula. In 1921 Robert A. Millikan, professor at the California Institute of Technology, famous for his experiments determining the charge of the electron, wrote a scathing criticism of engineering education in the United States to his friend Henry S. Pritchett. Millikan, who compared the U.S. system of education with the European, felt that too much emphasis was being placed on details of industrial processes and management. Others too were critical of engineering education, and the dissatisfaction led eventually to a comparative study, directed by William E. Wickenden and sponsored by the American Society for Engineering Education, of European and American engineering. Completed in 1929, the so-called Wickenden report concluded that U.S. engineering schools were not adequately training engineers to carry out engineering research and recommended better training, at both secondary and university levels, in mathematics and the basic sciences.

The Wickenden report, a new interest on the part of employers for engineers with graduate degrees, concern for allowing those employed full time to continue their education by means of evening courses, and the continued growth of radio engineering were some of the forces that made the years around 1930 turbulent ones for EE education. Nineteen hundred thirty was the year I joined the faculty of Brooklyn Polytechnic; the changes in EE education, as they appeared to me from that vantage point, are discussed next.

A Career in Electrical Engineering

V.1   From Vienna to Berlin to New York

As recounted earlier, I started work at the Vienna facility of Siemens-Schuckert in 1925. One of my colleagues was engaged in transformer and generator design and was particularly interested in higher-voltage phenomena. He called my attention to the difficulties of designing the shape of transformer windings and slots in the cores of generators so as to avoid high electrical field values and thus reduce the possibilities of breakdown and corona discharges, which were injurious to insulation materials. I found that my studies of function theory, especially conformal mapping, were useful in analyzing these problems. Since these mathematical techniques were not familiar to electrical engineers, I published several articles with my results in leading journals, including *Elektrotechnische Zeitschrift, Archiv für Elektrotechnik*, and *Elektrotechnik und Maschinenbau*.

Even though I was given interesting assignments—as, for example, designing an electric motor, for use in a mine, that had to have a terrific starting moment—I was not content to remain where I was. For one thing, the economic situation of Austria was not encouraging. For another, I knew of the outstanding work of the Siemens-Schuckert development division in Berlin, which was headed by Reinhold Rüdenberg. It proved easier than I expected to arrange my transfer, with the help of the director of the design division in Vienna, and at the beginning of January

1929 I began work at the Siemens headquarters in Berlin-Charlottenburg. My wife and I had found an apartment in a new suburban section of Berlin from where I would commute by electric trolley.

The economic and political situation in Germany was at the time quite troubled. The economy suffered from the need to pay war reparations and, more seriously, from the worldwide economic collapse following the U.S. stock market crash of October 1929. Political moderates seemed to be losing ground to extremists of the right and left. The death of the respected moderate Gustav Stresemann in October 1929 and the appointment of a Nazi cabinet member (Wilhelm Frick as minister of the interior) in January 1930 boded ill, I thought, for Germany's future.

I had hoped that by transferring to Berlin I would have the opportunity to become acquainted with some of the outstanding engineers then working in that city. One with whom I worked was E. Ollendorf, who later immigrated to the United States at my invitation. While in Berlin, I was invited to contribute to a conference, organized by engineers in Rüdenberg's development division of Siemens, devoted to the application of higher mathematics to engineering. The lectures given at this conference—including my lecture on "Fields near edges," an application of conformal mapping—were published in book form by Springer Verlag in 1932, and this book was translated into English and published by M.I.T Press as *Theory of Functions as Applied to Engineering Problems* (see Figure V.1).

Not long after I moved to Berlin, I was invited to give a paper at a meeting of the Vienna section of the electrotechnical association, no doubt because of the papers I had published on the analysis of electrical machines. I used the occasion of this trip to Vienna to step into the office of Professor Ehrenhaft, who had been advisor on my physics dissertation. Ehrenhaft wanted me to meet the newly appointed professor of telecommunications, Felix Petritsch, and phoned him at once, making an appointment for later in the morning. When I met Professor Petritsch, we talked for some time exchanging personal experiences. I talked about my work at Siemens, and he told me about a tour he had just completed of some technical universities in the United States.

THEORY OF FUNCTIONS

AS APPLIED TO
ENGINEERING PROBLEMS

EDITED BY

R. ROTHE, F. OLLENDORFF AND K. POHLHAUSEN

AUTHORIZED TRANSLATION BY

ALFRED HERZENBERG, DIPL. ING.

TECHNOLOGY PRESS
MASSACHUSETTS INSTITUTE OF TECHNOLOGY
CAMBRIDGE

1942

Figure V.1   The title page of a 1942 printing of *Theory of Functions as Applied to Engineering Problems,* originally published by M.I.T. Press in 1933.

About a half year later, I received a letter from Petritsch asking if I were interested in spending a year in the United States as a visiting professor of electrical engineering at Polytechnic Institute of Brooklyn. Of course I said, "yes," but wanted more details. He then sent me information and an application form. The appointment would be for September 1930 to June 1931 and would carry a salary of $5000. The successful applicant would teach two two-hour courses to graduate students, one in circuit theory and one in electromagnetic theory. I was very much interested in teaching and had in fact just been made *Privat Dozent* (academic lecturer) at the technical university in Charlottenburg.

After discussing the matter with my wife, I decided to apply. As a youngster I had read a great many of Karl May's novels set in the New World, and, as a result, I had long wanted to get to the United States. Unfortunately, the information specified that applicants must be at least 30 years of age, and I had not yet turned 29. Nevertheless, I thought there might be some hope, and I completed the application form, which asked many questions about previous experience, and even started taking classes at the Berlitz school in Berlin to learn conversational (British) English.

Very shortly afterward I was asked by the technical university in Charlottenburg to appear for examination for *Habilitation* (a degree beyond the Ph.D.), which I passed well because of my list of publications. Then in May a letter came from Polytechnic Institute telling me of my appointment as visiting professor and containing money for passage by boat from Bremen to New York.

My wife and I got busy with all preparations. We arranged for the care of our apartment and got visas for entry into the United States. I obtained a leave of absence from Siemens for one year. And I began preparing for the two courses I would teach.

Preparing for the course on electromagnetic theory was the easier task. I had thoroughly studied August Föppl's *Einführung in die Maxwellische Theorie der Electrizität* (1904), which was the first fairly complete presentation of Maxwell's theory to German audiences. (Föppl drew heavily upon—and fully acknowledged—Oliver Heaviside's development of Maxwell's theory.)

Another book I liked very much, and recommended to the class, was E. Bennett and H. M. Crothers's *Introductory Electrodynamics for Engineers* (1926). For the circuit theory course I prepared to teach Heaviside's operational calculus, which was at that time frequently taught and used in electrical engineering, particularly in the United States, where Ernst J. Berg promoted its use.

Near the end of the summer of 1930, my wife and I made the transatlantic crossing on the *Berengaria*, a luxury liner built in Germany in 1912 and originally called *Imperator*. In New York Dean Eric Hausmann of Polytechnic Institute met us and took us to a hotel in Brooklyn Heights. What I remember most about that hotel room is trying to cope there with the extremely hot and humid weather, helped only by a ceiling fan.

Classes at Polytechnic went better than I expected, even though I was not very fluent in English at the outset. The students were very polite and showed a great deal of interest, and they asked many more questions than was common in German classrooms. I found the teaching very much to my liking, so was delighted when, in February 1931, I was offered a permanent position at Polytechnic—as research professor of electrical engineering with an annual salary of $6000 (see Figure V.2). The appealing title and the higher salary made this attractive; also, the prospect of staying permanently in the United States appealed to me.

My wife and I returned to Austria that summer to obtain immigration papers for entry into the United States and to otherwise arrange for the transfer. While in Austria I happened to talk with a professor of the technical university, who urged me to get to know a certain Austrian family—Hugo and Sonya Eisenmenger and their two daughters—then living outside New York City. I had, of course, no inkling then that my subsequent visit to that family would change the course of my life.

V.2   POLYTECHNIC UNIVERSITY

In 1853 prominent citizens of Brooklyn Heights, most of them graduates of Ivy League colleges, founded Brooklyn Collegiate and Polytechnic Institute, a preparatory school and junior col-

Figure V.2 A picture taken at about the time I began work at Polytechnic Institute.

lege, to provide an educational institution for their sons close to home. It became a four-year institution in 1869 and changed its name to Polytechnic Institute of Brooklyn in 1889. At this time the four-year science course offered by the college was divided into three areas of specialization: chemistry, civil and mechanical engineering, and electrical engineering. The program in electrical engineering had been established in 1886, only a few years after the first such programs began at M.I.T. and Cornell.

In the 1890s and the first decade of this century a great wave of European immigration swelled the population of Brooklyn and lower Manhattan. Many young immigrants sought an engineering education as a route into society at a professional level. Few of them, however, had the means to attend school full time, so in 1904 Polytechnic Institute started an evening program that had the same requirements and standards as full-time day study. This meant that it took at least 8 years—often 10 or 12 years—to obtain a Bachelor's Degree. Thus a person could earn a professional degree while being employed full time, and thus Polytechnic could, in a new and very effective way, serve the needs of a larger community.

It was only gradually that employers in industry recognized the value of a graduate degree in electrical engineering. Some of the reasons graduate training became more valuable for practicing engineers were the advances in engineering science, the increasing complexity of power networks and electrical machines, and the rapid expansion of the new realm of radio engineering. Polytechnic was one of the first institutions in the country to offer an evening program for the Master's Degree. This began in 1926, four years before I arrived.

At that time the mechanical and civil engineering programs at Polytechnic were larger than the EE program. We gained many evening students in EE because of the proximity of employers such as Con Edison, Brooklyn Edison, Bell Labs, and Sperry Gyroscope. The two courses I taught the first semester were both in the evening EE program, and it was not long before I was placed in charge of that program. It grew rapidly, and in 1936 we added a doctoral program in electrical engineering.

My style of teaching seemed to be more conceptually oriented than the students were used to; I always avoided teaching just mathematical manipulations. My work at Siemens had shown me the usefulness of a theoretical understanding of electromagnetic phenomena, especially as embodied in Maxwell's equations, and I wanted my students to have the same advantage I had.

There was, however, the difficulty that the students entering Polytechnic had less mathematical training than entering students at German or French universities. Some years later I assisted the New York Board of Education in an appraisal of foreign diplomas; we found that a graduate of German secondary school—a recipient of the *Maturitätszeugnis*—had an advantage of almost two years of study over graduates of U.S. high schools. Fortunately at Polytechnic we had a strong faculty able to give rigorous training in mathematics and physics, and—what was most important—we had students who really wanted to learn.

My wife and I were divorced in 1933, and three years later I married the woman, Sonya Eisenmenger, whom I had met in the chance way described in the preceding section. Sonya was already well embarked on a successful career as a physiotherapist; in 1934 she received one of the first doctorate degrees in physiotherapy.

In the 1930s the graduate EE program at Polytechnic grew steadily. I laid out a program of courses in engineering science, technology, and management, and I began a research program myself in ultrahigh frequencies (up to 600 MHz), a region of the spectrum then little studied. With the coming of war in 1939, even higher frequencies became of great interest, principally because of their use in radar systems. I decided to focus my research on techniques for measuring frequencies, wavelengths, and power attenuation of microwaves (electromagnetic waves of frequencies from 3000 MHz to 300,000 MHz).

After a visit to the Massachusetts Institute of Technology shortly after the establishment of the Radiation Laboratory for the development of radar systems, I decided that the most pressing need was for a practical and accurate means of measuring the power output of a radar transmitter and the sensitivity of a radar receiver. For this purpose my colleagues and I at Polytechnic designed and built an accurately calibrated power attenuator. W. W. Hansen of Sperry Gyroscope had produced an attenuator consisting of a glass tube coated with Aquadag (a colloidal suspension of graphite, marketed as a lubricant), but this device was neither stable nor very accurate. I knew that a metal film, especially a noble metal film, would be much more stable. My colleagues and I designed a coaxial device, which contained, inside a metal housing, a glass rod coated with a platinum-palladium alloy, chosen for its very high resistivity.

We knew that the attenuator would be of great use to the military—for the calibration of radar sets in the field—only if it were extremely rugged, and we worked to achieve this quality. When Jerrold Zacharias of M.I.T.'s Radiation Laboratory examined the device, he tested its ruggedness in a simple way, by throwing it on the floor. When it did not break, he immediately ordered 1000 of them, and this order was shortly thereafter increased to 10,000.

I could see that it would be necessary to organize a company to produce the attenuators. I explained the situation to Dr. Harry S. Rogers, president of Polytechnic, and he and two trustees of Polytechnic each gave $10,000 for the initial capital. Thus PIB (Polytechnic Institute of Brooklyn) Products Company was founded. The company soon had contracts with all the military services and produced attenuators of various types. In 1946 its name was changed to Polytechnic Research and Development (PRD) Company, and the company's product line grew to include wave meters and other kinds of measuring instruments. (See Figure V.3.)

PRD continued to grow after the war, and its annual sales reached $5 million in the mid-1950s. From its founding I was director of the company, and in 1952 I became also president. Running the company was taking more and more of my time, and I realized that I had to make a choice between industry and academia. Though I had begun my career in industry, my heart was in academia, so in 1959 we sold PRD to Harris-Intertype (now Harris Corporation). The proceeds from this sale provided the first substantial endowment for Polytechnic.

Early in the war I was invited to join the Radiation Laboratory, but declined on the grounds that the research and educational programs at Polytechnic were extremely important to the war effort. Just after the war, however, I did become a Rad Lab employee for a short time in order to write two chapters (on microwave measurements) in a volume of the famous Rad Lab series of texts. Since the Radiation Laboratory closed down shortly

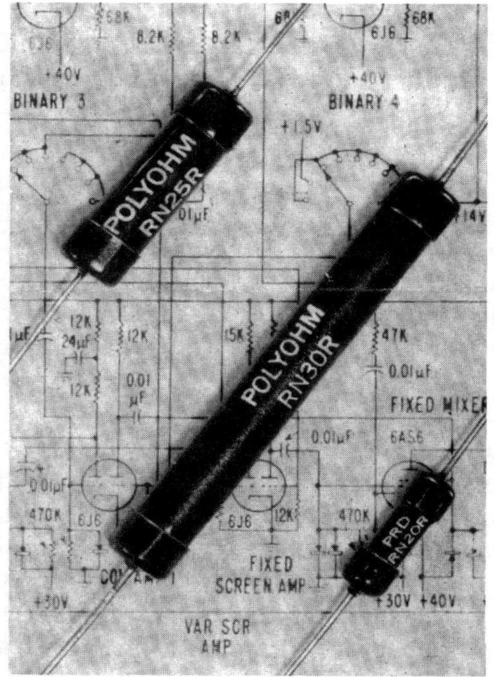

Figure V.3 Some polyohm resistors produced by Polytechnic Research and Development Company.

thereafter, it is possible that I have the distinction of being the last person hired by that institution.

After the war we established at Polytechnic the Microwave Research Institute (MRI), which became one of the most important centers for research into microwave theory and techniques (see Figure V.4). Beginning in 1952 we organized an annual conference, called the Microwave Symposium. Soon the three military services and the IRE joined MRI as sponsors of the conference. We were able to attract leading researchers, and the published proceedings served as compendia of recent work and became extremely influential. In later years, wherever I traveled in the world I would see these proceedings volumes.

In 1957, after a short period as vice president for research, I was named president of Polytechnic (see Figure V.5). One reason the Board of Trustees selected me was that I had been on the faculty for so long—28 years at that point—and worked well with

Figure V.4 Here an MRI colleague and I are observing measurements of the transmission of high-frequency waves in a waveguide.

Figure V.5 My investiture as president of Polytechnic Institute of Brooklyn by Preston R. Basset, chairman of the Board of Trustees.

other faculty members. A major regret about accepting the presidency was that I would not then be able to teach, and I had always found teaching very rewarding. (I did in fact teach one course for the first two years, but found that university and professional meetings and necessary travel so much interfered with the lecturing that I could not continue teaching.)

Because of the rapid growth of Polytechnic's programs, we were very crowded in the existing facilities. One of the first tasks I undertook as president was to direct the move to a new campus on Jay Street in Brooklyn. Then in 1961 we opened a new graduate center in Farmingdale, Long Island (see Figure V.6). The proximity of a number of high-tech companies was a principal consideration in that choice of location; it allowed many forms of cooperation, such as continuing professional studies and collaborative research, between companies and the university. (The EE department at Polytechnic began offering off-campus graduate courses for credit in 1942, and in that year I taught linear transient analysis at the Sperry Gyroscope Company.)

After the addition of the Farmingdale campus, we had two complete, parallel programs for graduate degrees in EE. Indeed, I saw the expansion of graduate programs, together with the strengthening of Polytechnic as a research institution, as the principal mission of my presidency. We achieved success at this: in 1970, the year following my retirement as president, 470 master's degrees were awarded, compared to 150 in 1958, and the number of doctoral degrees awarded had increased from 36 to 108.

Another principal task I set myself as president of Polytechnic was to better the financial health of the institution through fund-raising. Polytechnic had never done much fund-raising, and one of the first things I did was to institute that through the Alumni Fund. As mentioned, the sale of PRD in 1959 provided the school with its first substantial endowment, and, hoping to build on this start, I appointed a director of development.

Retirement from Polytechnic in 1969 left me more time for professional activities, which is the subject of the next section, but did not sever my ties to the institution as I continued to serve on the Board of Trustees. In 1973 Polytechnic merged with New York University School of Engineering and changed its name to Polytechnic Institute of New York, and ten years later

Figure V.6   The groundbreaking ceremony for the new campus in Farmingdale, Long Island.

the name was changed again to Polytechnic University. I was greatly honored in 1986 when Polytechnic's Microwave Research Institute was rededicated as the Weber Research Institute (see Figure V.7).

V.3   PROFESSIONAL ACTIVITIES

From my student days in Vienna I have been conscious of the importance of professional societies. Through publications, confer-

Figure V.7 The cover of the program for the dedication of the Weber Research Institute.

ences, and meetings they stimulate individual achievement and advance the profession as a whole. In 1923 I joined the Austrian Association of Electrical Engineers (*Österreichischer Verband für Electrotechnik*), and after I moved to Berlin in 1929 I became active in the Commission on Units and Standards (*Ausschuss für Einheiten und Formelgrössen*). (This interest in standards of measurement continued throughout my career, and I published several articles in this area, as well as authoring the section "Physical units and standards" in *Handbook of Engineering Fundamentals* (John Wiley, New York; 1936 and 1952).)

As soon as I came to this country I joined the American Institute of Electrical Engineers and the American Physical Society. In 1937 I organized the Basic Science Group of the AIEE. We arranged for lectures, held at the 39th Street engineering building, on the physics underlying electrical engineering. For example, in November and December 1941 we presented six lectures on nonlinear circuit theory, and in the following year the topic was ultrashort electromagnetic waves.

I was always attracted by the Institute of Radio Engineers (IRE). Their publications were usually physics oriented, and I regard physics as the basis of all electrical engineering. Also, the IRE from its beginnings was an international organization, and part of the reason for this is that Europe had no comparable journal, that is, one treating radio engineering as applied physics.

In 1959 I was president of the IRE. In that year there was a great deal of interest in science and technology—these were the early days of the satellite era—and I traveled throughout this country and abroad to give lectures. Also in that year there was some talk of bringing the IRE and the AIEE together in a single society.

The founders of the IRE had considered becoming a part of the well-established AIEE, but felt that their interests would be greatly overshadowed by those of power engineers and telegraph and telephone engineers. Moreover, the IRE founders wanted a society oriented toward science rather than one, like the AIEE, oriented toward industry applications.

Over the next four or five decades the two societies, for the most part, went their own ways. There were, though, areas of overlap, such as power electronics and engineering in medicine and biology, and there were many instances of cooperation, especially in setting standards. It became common in the years immediately following World War II for a college campus to have a common student branch of the two societies, and in 1950 both the AIEE and the IRE authorized such common branches.

The postwar world saw a remarkable growth and diversification of electronics engineering, and IRE's membership grew rapidly. At the same time, electronics became increasingly important in the traditional AIEE areas: power engineering and wire communications. My own view was that all areas of electrical and electronic engineering were based on the same underlying physics. It was also clear that a single society would be more efficient by eliminating duplications of staff, publications, and activities.

Therefore I was pleased to be part of a study group formed by the AIEE in 1961 to consider the question of whether the societies should merge. We unanimously recommended merger. That same year a joint AIEE-IRE committee was formed to consider the specifics of merger, and the following year the question was put to the membership of the two societies. Of those voting, 87 percent approved the merger, and on January 1, 1963, the Institute of Electrical and Electronics Engineers was officially established (see Figure V.8).

Figure V.8 The logos of the AIEE and the IRE at the time of the merger, and the logo of the newly formed IEEE.

It was a great honor to be selected by the joint merger committee to be IEEE's first president. (All subsequent presidents have been elected by vote of the members.) In my year as president I traveled a great deal, visiting about 40 of the 120 IEEE sections (see Figure V.9); I wanted to strengthen the feeling that the new organization would be successful and to do what I could to resolve any difficulties resulting from the merger. Also in that year I traveled abroad to stimulate interest in forming IEEE sections.

There were other professional organizations that I was privileged to be associated with: the New York Electrical Society (for which I served a term as president), the American Mathematical Society, the American Association of University Professors, the New York Academy of Sciences, and the Engineers' Council for Professional Development (where I was on the Board of Directors from 1964 to 1972 and was president from 1968 to 1970). In 1950 I became a member of the International Union of Radio Science (URSI), and I am still a member. Much of my work with URSI has been concerned with standards of measurement.

In 1935 I joined the American Society for Engineering Education. Polytechnic faculty members have been active in this society from its very beginnings. Indeed, its first meeting, held in 1874—it was then called the Society for the Promotion of Engineering Education—took place at Polytechnic. And Polytechnic has been quite innovative in its programs: it was one of the first colleges to offer a course of study in electrical engineering; also one of the first to offer a program of evening studies; and its ties to industry, such as off-campus courses, were emulated by other colleges.

In looking back on my life of engineering research, college teaching, university administration, and professional activities, I count myself extremely fortunate. Though I have almost always been extremely busy, I have not experienced my activities as a burden. I've always focused on the marvelous things that might be achieved—with a theory, with a student, or with an educational or professional organization—and found joy in my efforts to bring some of these possibilities to fruition.

Figure V.9 After attending a meeting of the Hudson Valley Joint Technical Societies in March 1963, I visited the IBM facilities in Poughkeepsie, New York, where I looked at the new IBM 7080 computer.

Notes on Sources

An extremely useful bibliography is Bernard S. Finn's *The History of Electrical Technology: An Annotated Bibliography* (Garland, New York and London; 1991). There are a few general histories of electrical engineering, notably W. A. Atherton's *From Compass to Computer* (San Francisco Press, San Francisco; 1984), P. Dunsheath's *A History of Electrical Engineering* (Faber and Faber, London; 1962), and John D. Ryder and Donald G. Fink's *Engineers and Electrons: A Century of Electrical Progress* (IEEE Press, New York; 1984). Some other general and useful books are Thomas P. Hughes's *American Genesis: A Century of Invention and Technological Enthusiasm*, 1870–1970 (Viking, New York; 1989) and Charles Susskind's *Twenty-five Engineers and Inventors* (San Francisco Press, San Francisco; 1976).

For topics covered in Chapter II, Section II.1, I recommend I. Bernard Cohen's *Benjamin Franklin's Experiments* (Harvard University Press, Cambridge, MA; 1941), John L. Heilbron's *Electricity in the 17th and 18th Centuries: A Study of Early Modern Physics* (University of California Press, Berkeley; 1979), and Duane Roller and Duane H. D. Roller's article "The development of the concept of electric charge, electricity from the Greeks to Coulomb," in *Harvard Case Histories in Experimental Science* (Harvard University Press, Cambridge, MA; 1954). An English translation of William Gilbert's *De magnete* was published by Basic Books (New York) in 1958.

For the next section one may profitably consult Ronald Kline's article "Science and engineering theory in the invention and development of the induction motor, 1880–1900," in *Technology and Culture*, Vol. 28 (1987), pp. 283–313, and Malcolm MacLaren's *The Rise of the Electrical Industry During the Nineteenth Century* (Princeton University Press, Princeton, NJ; 1943).

Biographies of Michael Faraday have been written by John Tyndall (*Faraday as a Discoverer*, Longmans, Green & Co., New York; 1868) and by L. P. Williams (*Michael Faraday, A Biography*, Basic Books, New York; 1964).

Joseph Henry's contribution to electrical technology is considered in Arthur Molella's article "The electric motor, the telegraph and Joseph Henry's theory of technological progress," in *Proceedings of the IEEE*, Vol. 69 (1976), pp. 1273–1275; Thomas Coulson has written a biography of Henry entitled *Joseph Henry, His Life and Work* (Princeton University Press, Princeton, NJ; 1950). The papers of Joseph Henry are being edited and published by the Smithsonian Institution. Submarine telegraphy is the subject of Charles Bright's *Submarine Telegraphs, Their History, Construction, and Working* (Crosby Lockwood, London; 1898).

A standard history of telephony is John Brooks's *The Telephone, the First Hundred Years* (Harper & Row, New York; 1975). Another rich source is E. J. Houston and A. E. Kennelly's *The Electric Telephone*, 2nd ed. (McGraw-Hill, New York; 1906).

Sources for Section II.6 include A. Bright's *The Electric Lamp Industry*, Macmillan, New York; 1949), I. C. R. Byatt's *The British Electrical Industry, 1875–1914* (Clarendon Press, Oxford; 1979), John A. Fleming's *Fifty Years of Electricity* (Wireless Press, London; 1921), Robert Friedel and Paul Israel's *Edison's Electric Light: Biography of an Invention* (Rutgers University Press, New Brunswick, NJ; 1986), and J. W. Hammond's *Men and Volts: The Story of General Electric* (J. B. Lippincott, New York; 1941).

For Section II.7, I made much use of Paul J. Nahin's *Oliver Heaviside: Sage in Solitude* (IEEE Press, New York; 1987) and M. Rowbottom and C. Susskind's *Electricity and Medicine: History of their Interaction* (San Francisco Press, San Francisco; 1984).

For the subjects considered in the final section of Chapter II, excellent sources are Philip Alger and Robert E. Arnold's article, "The history of the induction motor in America," in *Proceedings of the IEEE*, Vol. 64 (1976), pp. 1380–1383; Robert Belfield's article, "The Niagara System: The evolution of an electric power complex at Niagara Falls, 1883–1896," in *Proceedings of the IEEE*, Vol. 64 (1976), pp. 1344–1450; John A. Fleming's *The Alternating Current Transformer in Theory and Practice* (Electrician Printing and Publishing, London; 1889); John Hopkinson's *Original Papers on Dynamo-Machinery and Allied Subjects* (J. W. Johnston, New York; 1893); and Malcolm MacLaren's *The Rise of the Electrical Industry During the Nineteenth Century* (Princeton University Press, Princeton, NJ; 1943).

The principal sources for Chapter III are Christa Jungnickel and Russell McCormmach's *Intellectual Mastery of Nature: Theoretical Physics from Ohm to Einstein, Vol. I, The Torch of Mathematics, 1800–1870*, and Vol. II, *The Now Mighty Theoretical Physics, 1870–1925* (University of Chicago Press, Chicago; 1986) and L. Campbell and W. Garnett's *The Life of James Clerk Maxwell* (Macmillan, London; 1882). A reprint of Newton's *Philosophiae Naturalis Principia Mathematica* is available from Dover Publications (Mineola, NY), as is a reprint of James Clerk Maxwell's *Treatise on Electricity and Magnetism*. In 1890 Cambridge University Press published *The Scientific Papers of James Clerk Maxwell*.

The principal sources for Section IV.1 were Paul J. Nahin's *Oliver Heaviside: Sage in Solitude* (IEEE Press, New York; 1987) and the Dover reprint of Heaviside's *Electromagnetic Theory*.

Excellent sources for the history of EE education are A. Michal McMahon's *The Making of a Profession: A Century of Electrical Engineering in America* (IEEE Press, New York; 1984); Robert Rosenberg's "The origins of EE education: A matter of degree," in *IEEE Spectrum*, Vol. 21, July 1984, pp. 60–68; and Frederick Terman's "A brief history of electrical engineering education," in *Proceedings of the IEEE*, Vol. 64 (1976), pp. 1399–1407.

Heinrich Hertz's *Untersuchungen über die Ausbreitung der elektrischen Kraft* (Johann Barth, Leipzig; 1892) is available in English translation in a Dover reprint.

Sources for Section IV.4 include Hugh H. J. Aitken's *Syntony and Spark: The Origins of Radio* (John Wiley, New York; 1976) and Friedrich Kurylo and Charles Susskind's *Ferdinand Braun* (MIT Press, Cambridge, MA; 1981).

For more information on topics considered in the following section, see Thomas P. Hughes's *Networks of Power: Electrification in Western Society, 1880–1930* (Johns Hopkins, Baltimore and London; 1983) and James Brittain's article "The introduction of the loading coil: George A. Campbell and Michael I. Pupin," in *Technology and Culture*, Vol. 11 (1970), pp. 36–57.

An excellent source for the history of physics in the early twentieth century is Louis de Broglie's *The Revolution in Physics: A Survey of Quanta for the Layman* (Noonday Press; 1953).

Sources for the final section of Chapter IV include Hugh G. J. Aitken's *The Continuous Wave: Technology and American Radio, 1900–1932* (Princeton University Press, Princeton, NJ; 1985), W. R. Maclaurin's *Invention and Innovation in the Radio Industry* (Macmillan, New York; 1949), and Gerald F. J. Tyne, *The Saga of the Vacuum Tube* (Howard W. Sams, Indianapolis, IN; 1977).

A short history of Polytechnic University is contained in George Bugliarello's *Towards the Technological University: The Story of Polytechnic Institute of New York* (Newcomen Society in North America, New York; 1975).

Afterword

To have enjoyed the privilege of having been a student and later a colleague and friend of Dr. Ernst Weber can only be described as a superb, enriching experience. Many occasions of great enlightenment have highlighted my association with Dr. Weber, very similar to the power of magical musical moments that lifts one to greater heights of human emotion.

In fact, among his many cultural interests, Dr. Weber is a profound devotee of classical music. To observe Dr. Weber's eyes sparkle at a concert or opera as his emotions are stirred attests to his deep sensitivities. These sensitivities have guided him to enormous accomplishments with great humility. Dr. Weber is a person who inspires respect. He does not have the need to be overbearing or cynical. He speaks softly, and his thoughts are reflective of a wealth of knowledge realistically synthesized in achieving a better understanding of our world and plausible future directions. In another life, Dr. Weber could have been a wonderful musical director, because of his unique ability to select people capable of achieving a common goal with an unusual degree of harmony.

The autobiographical chapters of the text provide rare glimpses of Dr. Weber. His growth takes us through difficult times, in contrast to his professional life, which takes us through challenging times. One observes how personal plans are often altered by the circumstances of life beyond one's control. Throughout these historical steps one senses that Dr. Weber is endowed with a tremendous gift of optimism and appreciation for the many opportunities that became available as personal challenges.

Dr. Weber's retrospective reaction concerning these opportunities is expressed so well in Chapter V, Section 3, where he writes,

> In looking back on my life of engineering research, college teaching, university administration, and professional activities, I count myself extremely fortunate. Though I have been extremely busy, I have not experienced my activities as a burden. I've always focused on the marvelous things that might be achieved—with a theory, with a student, or with an educational or professional organization—and found joy in my efforts to bring some of those possibilities to fruition.

Early Years

Dr. Weber was born in Vienna in 1901, the oldest of five children. His main interest was philosophy, but dire circumstances caused him to think in terms of making a decent living. Consequently, he pursued a dual course of studies, simultaneously attending the Technical University of Vienna for engineering and then tramping across town to attend the University of Vienna for philosophy, physics, and mathematics.

The expense of the actual schooling was nominal, perhaps $50 a year for everything. But still there was sacrifice. His early years spanned World War I, and hunger was widespread in Vienna. Dr. Weber would tutor the baker's son in return for a loaf of bread. His family was not well-to-do. Still, despite the miserably bad times, he persisted in getting an education.

After receiving his Diploma of Engineering in 1924 from the Technical University, he joined the Austrian Siemens-Schuckert Company as a research engineer. It was the equivalent of a General Electric or a Westinghouse. Dr. Weber received his Ph.D. from the University of Vienna in 1926 and the D.Sc. from the Technical University in 1927. In 1929, he transferred to the Siemens-Schuckert in Berlin to concentrate on advanced design. At the same time, he received an adjunct appointment as lecturer at the Technical University.

While lecturing at the Technical University, a colleague who had just returned from America advised him that the Polytechnic Institute of Brooklyn was seeking a visiting professor to initiate its graduate school in electrical engineering. He encouraged Dr. Weber to apply. In those days, the Massachusetts Institute of Technology had taken the lead in offering graduate programs.

Although Dr. Weber was a year younger than the required age, he applied and was accepted—much to his surprise. Thus began a new career in a different environment with most promising opportunities. In fact, the first year was highly successful, so much so that Polytechnic offered him its first research professorship and urged him to stay on.

Most of the early research concerned power engineering. Telephony was the big thing. AT&T was just building up. The focus was on multispeech transmissions. At the same time, ConEdison had its own research laboratory and would share some of its research problems with Dr. Weber and his students. The whole theory of electrical phenomena was still in the making. Dr. Weber organized basic courses in circuit systems and electromagnetic theory, which were offered for graduate credits toward an advanced degree.

Dr. Weber soon gained an esteemed reputation with his students. He had demonstrated a deep interest in developing each student to his maximum potential. His evening courses attracted students from major research centers, including Bell Laboratories and Sperry Gyroscope. They provided him with the latest developments and experimental products to update his research laboratory with state-of-the-art equipment.

Dr. Weber often met with his students for lunch. On one occasion, a student asked Dr. Weber, "What is life really all about?" Dr. Weber replied: "I believe life is an evolution, an ever-changing vast experiment, a mysterious and puzzling experiment, yes, but one in which each of us has a definite role to play as a part of the total design." If each of us has a role to play, Dr. Weber has chosen many paths in his own quest to find purpose and meaning to life.

Development of Graduate Programs

Dr. Weber has written widely on the development of graduate programs in this country. I should like to summarize some of his observations here.

In the years before 1910, there was hardly any graduate work in engineering in America. It was not needed. We had reasonable prosperity and no incentives to strain mental ability. Then came World War I and with it a number of challenges to this country.

Two events took place around that time in electrical engineering. One was the development of the radio, bringing with it emphasis upon electromagnetic theory. The other was operational analysis. Together they became the two mainstays of required courses in the graduate curriculums in electrical engineering.

To update practicing engineers with current information, evening graduate programs began to be established in the 1920s. Several schools took the initiative, including the Polytechnic Institute of Brooklyn in 1926, which attracted much interest in the metropolitan area.

World War II brought again a number of developments that molded our whole concept of education in electrical engineering. The most important new concept was feedback. All the principles of servomechanisms and automatic feedback central systems were developed during World War II. Also, the ideas of information theory were coined, and virtually all microwave and radar techniques had their origin in war requirements.

Research has always been at the heart of graduate studies regardless of the discipline. Originally, research was carried on by the faculty without overall coordination. It was a master-pupil relationship. However, through the establishment of laboratories such as the Radiation Laboratory at MIT during World War II, the concept of research sponsored by the military emerged. After the war, this desirable partnership continued at individual universities, which made research results available to the military, but left military obligations to the military laboratories. Also, industry favored sponsored research as mutually beneficial. However, the problem of organizing university research on a more systematic basis remains a challenge because of lack of stability as sponsors come and go.

Genesis of Sponsored Research in Microwaves

When the European war started in 1939, Polytechnic was already heavily involved in graduate studies in electrical engineering in ultrahigh frequencies (the range of 100 to 500 MHz). By 1941 Dr. Weber had organized a comprehensive program of graduate lecture and laboratory courses in "Ultrashort Electromagnetic Waves." The program was scheduled to be given under the auspices of the U.S. Engineering Defense Training Program, which began in December of that year.

During this time, the Radiation Laboratory at MIT was founded. Physicists and engineering professors who had studied electromagnetics were invited to join the Radiation Laboratory. A few went to other groups at Harvard and Johns Hopkins. Also, the term "microwaves" was censored and placed on the government secret list for national security reasons. The government didn't want to divulge that America was trying to build radar systems in very high frequencies.

Shortly after Pearl Harbor, on December 28, 1941, Dr. Weber was visited by Drs. I. I. Rabi and F. W. Loomis. An invitation was extended to Dr. Weber and his microwave group of research associates (composed of Anthony B. Giordano, John W. E. Griemsmann, Stanley A. Johnson, and Nathan Marcuvitz) to join the Radiation Laboratory at MIT.

Dr. Weber faced a difficult decision. The choice was either to carry through the graduate program in the ultrashort waves and to continue the local leadership in the new and vastly expanding field or to join the rapidly growing group at the Radiation Laboratory at MIT. The relative importance and magnitude of service to the country in the two alternatives was deliberated at great length. Dr. Weber finally decided to remain at Polytechnic to carry forward a vigorous graduate training program in ultrashort waves for updating the major industrial research and development laboratories of the metropolitan area. In addition, it was agreed to accept a research contract to enhance the wartime effect with the Office of Scientific Research and Development (OSRD) under the sponsorship of Division 14 of the National Defense and Research Council (NDRC) in an area of critical importance.

The wisdom of this decision was demonstrated by the fact that in the course of the years up to June 1944, more than 750 graduate students had enrolled in the "Ultrashort Electromagnetic Waves" graduate training program. In addition, the services under the OSRD contract were continued in increasing measure up to the final termination date of OSRD activities.

On February 1, 1942, preliminary discussions were initiated on the subject matter to be covered by the proposed OSRD contract. A decision was made by Dr. Weber to concentrate upon the development of microwave radar yield and laboratory test equipment and components, both of which were in critical demand. Shortly after the contract began, Dr. Marcuvitz joined the Radiation Laboratory at MIT upon a special invitation extended through Dr. Weber.

The major contribution of the OSRD contract focused upon the development of accurate, reliable microwave attenuators by means of special techniques of precision metallization of glass perfected for this specific application. This brought with it a fundamental exploration of power attenuation measurements in the frequency ranges from 750 to 25,000 MHz, the design of attenuation standards, and the design of many basic test components and accessories for precision microwave measurements in the frequencies. It also led into an extended study of radio frequency (RF) fittings, in particular cable fittings, resulting in new designs for uses where precision was important.

Under the pressure of demand for the metallized glass attenuators from the armed services, it became necessary to organize a test and calibration company. It proved not possible to interest existing companies in the exacting and costly calibration phases of test equipment production. This company, PIB Products, Inc., together with Corning Glass Works as a subcontractor for the metallized glass elements, produced and calibrated more than 12,000 coaxial attenuator inserts of all sizes for fixed and variable types and delivered more than 11,000 metallized glass plates for use in waveguide attenuators. About 1450 of these plates were installed in attenuator casings with precision mechanical drive mechanisms and calibrated over specified frequency bands and ranges of attenuation. Continuous guidance on methods, consultation on schedules, and assistance with equipment and personnel were provided to both PIB Products and Corning Glass Works. PIB Products began operations in 1944 with William A. Lynch as manager and Harry C. Nelson as test facilities and measurement specialist.

The announcement of V-J Day initiated contacts with various branches of the services that were interested in the continuance of certain phases of the OSRD contract. The services included the Army, Air Forces, the Navy Bureaus of Aeronautics and of Ships, the Signal Corps, the Aircraft Radiation Laboratory, the Naval Research Laboratory, the Watson laboratories, and other governmental as well as industrial research laboratories.

Much interest was expressed, and the original "microwave research group" was reorganized in 1945 as the Microwave Research Institute with Dr. Weber as director to begin a new phase of microwave research with several services. PIB Products, Inc., became the Polytechnic Research and Development Corporation (PRD) to specialize in industrial products in the microwave range. Additional details are provided in Chapter V, Section 2, by Dr. Weber elaborating upon the leadership role of the Microwave Research Institute as well as subsequent developments concerning the company, PRD, which he organized and directed.

Contributions as President of Polytechnic

Dr. Weber's appointment as director of the Microwave Research Institute set the stage for a reorganization of the Department of Electrical Engineering where he served as head during 1945–1956. Under his leadership, enrollment in electrical engineering grew to represent 38 percent of the Institute enrollment of 5500 graduate and undergraduate students. The graduate program in electrical engineering developed into one of the largest in the country and soon gained a national ranking of sixth in its quality according to a study of graduate education undertaken by the American Council on Education.

In 1956 Dr. Weber was the first person to fill the new position of vice president of research. Following the untimely death of President Harry S. Rogers on June 6, 1957, he was appointed acting president. Six months later, Dr. Weber was appointed president and was inaugurated in the spring of 1958 as the sixth president of Polytechnic.

Prior to Dr. Weber's presidency, much planning had been accomplished under the leadership of Dr. Rogers from 1940 to 1955 and later during the centennial year of Polytechnic, which was founded in 1854. In fact, Polytechnic faced its hundredth birthday with the largest student enrollment in its history, a vigorous and expanding research program, and a sizable faculty whose reputation had spread worldwide. It was now time to consider new quarters for the burgeoning Polytechnic. The centennial provided such an opportunity.

A building fund was initiated and a search began for a suitable site for the new Polytechnic. Fortunately, the American Safety Razor Company had just vacated its block-square industrial plant facing Jay Street in Brooklyn and had moved its operation to the South. With imagination and great daring Dr. Rogers prevailed upon the trustees to buy the property for $2.1 million. The transfer was effected, and the property was on its way to becoming the new home for Polytechnic, which had survived for a century in quarters on Livingston Street and in a multitude of surrounding rented quarters.

The renovation of the new site had progressed far enough by the spring of 1957 that plans were made for the transfer of almost one-half the laboratory operations and all the classroom schedules to the new location. Optimism prevailed, but suddenly tragedy struck with the demise of Dr. Rogers, who had guided the destiny of Polytechnic with great promise for 24 years. It was now the responsibility of Dr. Weber to facilitate the completion of the new campus and to continue the expansion of educational programs in preparation for the future.

As a first step, Dr. Weber set a schedule for the completion of the conversion of the eight-story building on Jay Street as well as the five-story administrative building at the corner of Jay and Johnson Streets. Next, he devised a new table of organization, which was approved by the corporation in early 1958. It led to an amazing array of changes to promote a team effort. The faculty and staff rallied in a most gratifying way to meet the challenges, to master all the exigencies, and to continue the effort of building a new Polytechnic.

On April 19, 1958, Dr. Weber was inaugurated as the sixth president of Polytechnic, an occasion that was accompanied by the dedication of the eight-story building as Rogers Hall. It was a impressive ceremony with almost a thousand persons in attendance from all over the world. Here, before the converted industrial plant, the then Rear Admiral Hyman G. Rickover hailed Polytechnic as "an institution which has always been distinguished by single-minded concern with matters of the mind, and which can match its atmosphere of scholarly austerity with that of any of Europe's famous and remarkable centers of learning."

Thus began Dr. Weber's years as president—with much hope, trepidation, and wondrous concern in exercising the rights, privilege, and the responsibilities of his office. As splashes of paint began to be applied to complete the remarkable renovation of the Jay Street city block, Dr. Weber took stock of the situation at Polytechnic and concluded that it was in a strong position to move ahead with a faculty fully committed to the concept of scientific engineering and devoted to evolving new courses and curricula, contributing research papers, and encouraging initiative in the humanities and social sciences. It became clear that Polytechnic was very well prepared to assume the responsibilities of a broad program of quality engineering education at all levels coupled with vigorous research programs of interdepartmental character as preparation for leadership.

Dr. Weber began his years as president with an ambitious program to implement new developments in serving the community at large, including Long Island, as well as the nation and the world. He envisioned the ratio of doctoral to first degrees awarded by Polytechnic increasing manyfold, with consequent reaction on the undergraduate programs, making them truly oriented toward modern developments. This would require strong measures to further strengthen the departments at Polytechnic in teaching and research. He further envisioned Polytechnic gaining national prominence as a technological university through leadership in programs leading to the M.S. and Ph.D. degrees in science and engineering.

It is interesting to note that the largest number of degrees in Polytechnic's history was awarded at commencement in June 1970, following the retirement of Dr. Weber: the M.S. awards increased from 150 (in 1958) to 470 (in 1970); the Ph.D. awards increased from 36 (in 1958) to 108 (in 1970); the B.S. awards remained relatively constant. In 1958, there were 7 graduate programs leading to the Ph.D. degree. That number increased to 16 in 1970.

Among the many innovations and achievements that were accomplished during Dr. Weber's tenure as president are the following:

- A common freshman year for all students at Polytechnic was introduced to improve preparation for upper-level courses.

- The experimental undergraduate honors program in electrical engineering was expanded into a comprehensive interdepartmental unified honors program for especially gifted students entering their junior year. In June 1960 the Ford Foundation granted to Polytechnic a sum of $700,000 for the support of the honors program in science and engineering over a period of five to seven years.

- With the intense interest in the fields of high-power microwaves and plasma studies that could not be housed in downtown Brooklyn, grew the concept of the Polytechnic Graduate Center for graduate studies and research. A search began for an appropriate site that would combine the ongoing off-campus program in graduate studies being offered in Mineola for industrial personnel. The search narrowed to the Nassau-Suffolk county line as the probable future center of gravity of the industrial complex on Long Island. Republic Aviation Corporation came to the rescue by deeding to Polytechnic 25 acres of land in Farmingdale ideally suitable for the proposed Long Island center. Mundy I. Peale, Republic's president, called this "an investment in the future of the entire Long Island community." Ground was broken in the spring of 1960, and classes began in September 1961. The center grew to 1200 graduate students within a few years, offering advanced studies in science and engineering to the Long Island industrial community. The center stimulated close cooperation between Polytechnic's faculty and industrial research laboratories in areas of mutual interests.

- To keep abreast of an ever-changing curriculum, the faculty was encouraged through supporting services to write and publish text material internally; later many of these publications were published by McGraw-Hill as textbooks in a special series.

- An Industry Research Associate Program was initiated to foster cooperation between industry and education; many seminars were sponsored in specialized areas on emerging technologies.

- A new graduate program in electrophysics was introduced leading to the M.S. and Ph.D. degrees. It was the first of its kind, integrating electromagnetics, electrophoresis, and microwave networks and technology.

- A computer center was established with the latest computers to enhance the education and research efforts.

- A special office for continuing professional studies was organized for young graduates to keep abreast of advancing technology, and a "Modern Engineering Series" was introduced, as well as short courses in special techniques.

- The move in 1957 from the old Livingston Street complex to the Jay Street location was the first step in the expansion program. However, it was known even then that additional facilities would be needed as we became landlocked on Jay Street. To correct this, containing efforts would be necessary to acquire part of the surrounding area from urban renewal. The Brooklyn Polytechnic Urban Renewal project was created by the city council. Unfortunately, the project was abandoned when the new city administration selected another area in dire circumstances as the highest priority in urban renewal.

- Academic planning continued with a proposal to offer a premedical college program that gave recognition to the growing need for broader scientific preparation for medical study.

- A need developed to focus upon urban problems as part of the social change in the application of technology. This led to the formation of CUES, the Center for Urban Environmental Studies, to confront urban problems in transportation, control of air pollution, urban information communication systems, and optimum organization of city departments.

- In the area of biomedical engineering, stress was placed on the practice of medicine in the urban community and the offering of special courses in the local hospitals.

- Plans were set in motion for short courses in "Elements of Management" to meet the needs of technical personnel in managerial positions.

- A major addition to the physical plant was the Graduate Student Residence of the Long Island Graduate Center, which was the first dormitory facility in the Institute's 110-year history.

- Polytechnic gained an important foothold in the tract designated by the city of New York as the Brooklyn Polytechnic Urban Renewal Area; it was accomplished by the acquisition of a three-story building near the Jay Street campus.

- The corporation accepted a ten-year plan that was proposed by an ad hoc committee on planning for Polytechnic's evolution into a technological university. The plan was greeted with enthusiasm by all. Admittedly, the goals were set very high, and it would take tremendous effort to achieve them. The first step called for the creation of these academic divisions: Division of Sciences, Division of Engineering, and Division of Humanities and Social Sciences, each with a dean to provide leadership.

- The National Science Foundation (NSF) awarded a grant on November 3, 1965, in excess of $3 million for a three-year period to establish at Polytechnic "A Center of Excellence in Chemistry and Electronics." The grant brought to a climax one aspect of the academic and physical planning for the ten-year period 1963–1973.

- Faced with the urgent need for space, a decision was made to consider an 18-story academic tower on the Jay Street campus if no other option became available.

- A major laboratory was built and dedicated as the Preston R. Bassett Research Laboratory in April 1966 at the Long Island Graduate Center to replace facilities formerly at Freeport, Long Island. Part of the new facility included supersonic and hypersonic wind tunnels and other equipment for research in high-speed flight dynamics, interplanetary communications, and electrophysics.

- The national headquarters of the Engineering Concepts Curriculum Projects (ECCP) was transferred to Polytechnic in the fall of 1967 under the support of NSF. The ECCP was organized to generate material for a high school course presenting the impact of technology on society. An ECCP text, "The Man-Made World," was developed and used in over 60 high schools across the nation.

- The Carter Report (undertaken by the American Council on Education) entitled "Assessment of Graduate Education" and issued in 1967 placed Polytechnic's doctoral programs into the leading category nationally.

- At the conclusion of five years of the Engineering Concepts Curriculum Project, "The Man-Made World" was being offered in 400 high schools in 1969 (to 15,000 students) and in 20 liberal arts colleges.

Dr. Weber retired at a crucial time when plans were being considered to merge Polytechnic into the State University of New York within the next five years as the Technological University of New York. A number of proposals were prepared concerned with optimum plans. However, other circumstances prevailed, and Polytechnic remained a private university seeking its own destiny. Dr. Weber had paved the road to future growth in an urban environment as a unique resource recognized for high-quality faculty and programs.

Dr. Weber was often sought for advice in matters concerning professional development. In an article in the October 11, 1965 issue of *Electronic News,* Dr. Weber wrote:

> Engineering study by itself will not guarantee success in professional life; it can give only a foundation upon which each individual must build his own life guided, if he seeks it, by the advice of trusted elders. I would want to emphasize again, however, that the individual performance must primarily be attributed to the individual's own choice in accordance with his abilities.

Again, in his commencement address at Polytechnic in June 1969, just two weeks before his retirement, Dr. Weber expressed his opinions on this theme of individual worth. He said:

> True democracy needs the complete diversity of services from the most humble to the most demanding, physically and mentally. Rather than preach unreal and unachievable equality, we must

restore dignity to every phase of human endeavor from the laborer to the highly skilled craftsman…from the humblest draftsman to the most creative scientist and engineer.

Mrs. Weber: The President's Lady

Dr. Weber's marriage to Charlotte Sonya née Escherich, also a native of Vienna, provided constant support and loving care. They met by a chance introduction in America, and their marriage, which occurred in 1936, inspired much admiration. She was the daughter of Theodor Escherich, a pediatrician and bacteriologist famous for his discovery of the intestinal bacteria now named after him, Escherichia coli (E. coli).

Sonya Weber became well known as a physiotherapist—she had earned a doctorate in physiotherapy in 1934—and was for many years a director of the children's clinic of the Columbia-Presbyterian Hospital in New York City. She established an outstanding reputation and was in great demand as a lecturer on physical fitness. She and Dr. Krans developed the Krans-Weber physical fitness test used worldwide.

On the occasion of the establishment of the Sonya and Ernst Weber Scholarship Fund at Polytechnic in 1975, her son-in-law, James Flack, commented: "This Scholarship Fund in their names together is most fitting. They live and give of themselves as one. They have the same objectives, the same high standards. Each is the other's greatest supporter." Sonya Weber died in 1984 and is lovingly remembered in a book that Dr. Weber wrote about her life, family, and contributions.

Dr. Weber's home served as a mecca for students, friends, and colleagues. Their charming Tudor home in Mt. Vernon, New York, was the site of many happy occasions where people socialized and enjoyed themselves. Particularly memorable was viewing color slides of the Webers' travel experiences and listening to Mrs. Weber playing the piano.

Also, Mrs. Weber took a great interest in faculty wives by sponsoring special luncheons at a variety of city locations. She was very caring of the faculty's family health problems and provided personal assistance and advice whenever needed. She welcomed phone calls at any time.

During Dr. Weber's presidency, Mrs. Weber organized a series of yearly scholarship parties to honor individual countries that supplied graduate students to the international student body of Polytechnic. The goal was to raise funds to help those students to meet expenses. These were gala occasions attended by the ambassador and other officials of the country being honored as well as Polytechnic's faculty and staff and top managers of major companies. They are well remembered. Mrs. Weber earned a caring reputation by the letters she would personally write to solicit funds and special products from international countries being honored. The responses were usually very gratifying. Through checks received and the sale of donated products, the yearly scholarship effort amounted to around $50,000 in net funds.

Dr. Weber: The Teacher

Teaching was Dr. Weber's major interest in spite of efforts in other directions that were professionally related. He enjoyed the laboratory and classroom and direct interaction with students. In Chapter V, Section 2, Dr. Weber writes:

> My style of teaching seemed to be more conceptually oriented than the students were used to; I always avoided teaching just mathematical manipulations. My work at Siemens had shown me the usefulness of a theoretical understanding of electromagnetic phenomena, especially as embodied in Maxwell's equations, and I wanted my students to have the same advantage I had.

His style of teaching had a profound effect on the lives of his students, whether full time or part time. Students remember Dr. Weber coming into the classroom like a cool, refreshing breeze to provoke thought and ideas. He understood as no one else seemed to the enterprise of learning. One could feel the process of his thinking in contrast to the glib dramatics of the teacher-actor.

In his courses, the exchange of ideas was woven into the fiber of the subject. Dr. Weber knew what ideas were, where they belonged, and how to relate them to other ideas. Those classroom hours with Dr. Weber were thrilling. They aroused curiosity in an atmosphere that served as a stereoscope through which students could see thoughts taking on these dimensions.

Dr. Weber won his students by understanding and loving his subject as he understood and loved his students—always gentle, patient, kindly, and considerate, with the warmth of irrepressible eloquence. His publications have been many and are very well read. However, Dr. Weber wrote hundreds of other books by planting seeds in the minds of his students that later blossomed into many publications to enrich other intellectual lives.

One of Dr. Weber's principles of faith has been that the big things of life lie in the future, and consequently, he has always been looking ahead. No one can ask more than that of life—a chance to play a part in painting the dawn for the coming generations.

Dr. Weber has lived part of his life in the lives of his students. That is one reason why he discovered the fountain of perpetual youth. His thoughts are ever on the morrow. His eyes are turned continually toward the east where he knows the sun will rise.

Someone once inquired about Dr. Weber, "What does he teach?" One of his students responded proudly, "Well, he teaches many advanced subjects. But it's not what he teaches, the subject, I mean. The subject doesn't matter. It's what he is, the personality, and all that sort of thing. It's something he gives you, inspiration, new ideas, a fresh light on things in general. It's something he makes you want to do or be. I don't remember all the topics he taught, but I'll never forget that man—Dr. Ernst Weber, the teacher."

I, too, share the sentiment that I'll never forget Dr. Ernst Weber. It is with a sense of pride and respect that we salute you and wish you well, for a teacher like you affects eternity.

Anthony B. (Tony) Giordano